This fascinating book explains why the materials we can see and touch behave in the way that they do. In a completely non-technical style, using only basic arithmetic, the author explains how the properties of materials result from the way they are composed of atoms and why it is they have the properties they do: for example, why copper and rubies are coloured, why metals conduct heat better than glass does, why magnets attract an iron nail but not a brass pin, and how superconductors are able to conduct electricity without resistance.

The book is intended for general readers, and uses mainly words, pictures and analogies, with only a minimum of very simple mathematics. Even so, the author explains the basic ideas of quantum mechanics, the laws of which, although unfamiliar to most people, govern the behaviour of all matter in the universe. He explains how it is possible to understand the basic properties of matter using these ideas, and translates the technical jargon of physics into a language that can be understood by anyone with an interest in science who wants to know why the world around us behaves in the way that it does.

WHY THINGS ARE THE WAY THEY ARE

B. S. CHANDRASEKHAR

WHY THINGS ARE THE WAY THEY ARE

PUBLISHED BY THE PRESS SYNDICATE OF THE UNIVERSITY OF CAMBRIDGE
The Pitt Building, Trumpington Street, Cambridge CB2 1RP, United Kingdom

CAMBRIDGE UNIVERSITY PRESS
The Edinburgh Building, Cambridge CB2 2RU, United Kingdom
40 West 20th Street, New York, NY 10011–4211, USA
10 Stamford Road, Oakleigh, Melbourne 3166, Australia

First published 1998

Printed in the United Kingdom at the University Press, Cambridge

Typeset in Monotype Columbus 10½ on 13pt [WV]

A catalogue record for this book is available from the British Library

Library of Congress Cataloging-in-Publication Data
Chandrasekhar, B. S. (Bellur Sivaramiah), 1928–
Why things are the way they are / B.S. Chandrasekhar.
p. cm.
ISBN 0 521 45039 X (hc). – ISBN 0 521 45660 6 (pbk.)
1. Condensed matter. 2. Quantum theory. I. Title.
QC173.454.C49 1998
530.4'1–dc21 96–52937 CIP

ISBN 0 521 45039 X hardback
ISBN 0 521 45660 6 paperback

CONTENTS

PREFACE

A friend once remarked ruefully to me, 'Whenever I hear someone talk about a well-rounded person, I think of a spherical object with no distinguishing features on it.' A closer examination of such a person does reveal some features; these, however, are likely to be concentrated in areas that are labelled literature, art, music, philosophy, history and others generally called *the liberal arts* or *the humanities.* The area called *physics,* on the other hand, usually tends to be blank, although physics is as impressive an achievement of the mind as any of the others. A reason for this state of affairs is contained in the response I often get from someone who hears that I am a physicist: 'Physics was my most difficult subject in school. I never understood it.' Instead of giving up, I try to explain to the person some problem I am working on at the moment, maybe in superconductivity. I use words, analogies, pictures drawn on a paper serviette: anything but mathematics that is more advanced than simple arithmetic. I am then, even if not always, rewarded by the person with a dawning of interest and a desire to learn more. Having enjoyed this activity during all these years that I have myself been learning physics through teaching and research, I decided to gather and organize these ideas into a book for people (and I believe there are many) who are curious about the physicist's picture of the world, especially the part which forms our immediate surroundings.

There are many books which attempt to explain to a general readership certain parts of physics: stars and galaxies, quarks and neutrinos, the nature of space and time, the big bang, and relativity, to name a few. These involve ideas which have caught the public imagination and appear often in the science sections of newspapers and magazines. There is, however, a paucity of books which deal with that part of physics which is around us all the time and impinges on our daily lives, namely the physics of condensed matter. This is a branch of physics that explains the rich variety of properties displayed by the materials that make up the things that we can see, touch and use in our daily lives. It is a beautiful synthesis of three of the cornerstones of the subject, namely quantum mechanics, statistical physics, and electromagnetism, and is therefore an excellent means to discovering them. It is also the branch which continues to have the most applications to the development of objects used in everyday life: a fact which should please those who seek relevance in the endeavours of the scientist. This book is the result of my conviction that no one should be denied some knowledge of this fascinating field merely because of not being a physicist.

In writing this book, I have avoided certain features which are found in some books on popular science for the lay reader. To follow the line of reasoning in an unfamiliar field requires some effort, and I have left out things which seem to me irrelevant or distracting. I have drawn the figures so that they illustrate a point in the text, and nothing more. They are in effect the sketches on a serviette that I mentioned earlier. The reader will find no photographs of physicists or of equipment; it is not clear to me that looking at a picture of Enrico Fermi helps one to understand what a fermion is, or that gazing upon a photograph of tubes, cables, electronic instruments and assorted hardware clarifies the photoelectric effect. I describe the subject as it is today, and not its history. I name a few of the physicists who made signal contributions, but do not try to be exhaustive. I do not identify their nationalities or prizes won: physics is an endeavour which knows no national boundaries, and is not some kind of Olympics. I do not list books for further reading, because I think that readers whose interest is aroused will do much better browsing in libraries and bookshops and making their own discoveries. Two rules guided me in the selection of topics for inclusion in the book: they should show clearly how the basic concepts are applied, and they should be no more in number than could be included in a book of manageable size. In short, the book is lean, and intentionally so.

Several friends in fields like languages, law, history, classics, theatre and music, as well as physics, read parts of the manuscript and expressed their opinions, for which I thank them. I wrote the book while enjoying the hospitality and the environment of the Walther Meissner Institute of the Bayerische Akademie der Wissenschaften in Garching, Germany, and for this I am deeply grateful to its director, Professor Klaus Andres. To my wife Dorothee, who has helped me in more ways than she knows, I express my eternal gratitude.

B. S. CHANDRASEKHAR

GRÖBENZELL, GERMANY

1 INTRODUCTION

When you use a plastic spoon to stir hot coffee, the handle does not get hot, but if you use a silver spoon, it does. A sheet of glass is transparent so that you can see right through it, but a much thinner sheet of aluminium foil is opaque and lets no light through. You can bend a piece of copper wire back and forth without breaking it, but a glass rod is brittle and will break if you try to bend it. A magnet will attract an iron screw, but not a brass screw. A piece of copper is reddish in colour, while silver is – well, silvery. The electric current flowing through the tungsten filament of a light bulb makes it so hot that it gives off light. This current also flows through the connecting copper wires but does not make them even warm, and it cannot flow at all through the rubber insulation that covers the wires. The filament, the wires, and the insulation behave quite differently from one another as far as the electric current is concerned.

These examples illustrate the basic point I want to make: we are in our everyday life surrounded by objects that display a wide variety of properties. All these objects are made up of tiny things called atoms. A piece of silver is composed of silver atoms that are all exactly alike but are different from the atoms of, say, copper. Each atom consists of a nucleus (plural: nuclei) which is surrounded by a certain number of electrons. This number for silver is different from the one for copper, or for any other element. Each nucleus itself is made up of two kinds of even tinier things called protons and neutrons. The number of protons equals the number of electrons in the atom, and this number determines which element it is.

Thus all the different atoms, one for each of the approximately one hundred elements that exist in the universe, consist of just three kinds of objects: electrons, protons and neutrons. Yet when they are put together to give us the things we see around us, the result is a great range of different properties. One may well wonder how such complexity can result from putting together atoms composed of just three simple entities: electrons, protons and neutrons. Indeed, we shall see later that we need to consider only two entities, namely electrons and nuclei, to understand such properties of matter. That a nucleus is itself composed of protons and neutrons will turn out to be largely irrelevant for this purpose, though it is of crucial importance when we want to understand specifically nuclear properties such as radioactivity and nuclear fission.

Consider the silver spoon. It is made up of a very large number of silver atoms that must be stuck fast to one another, because the spoon retains its shape and does not collapse into a pile of free atoms. What is the 'glue' that holds the atoms together? Each atom is a piece of matter and weighs something, however little: it has mass. There is therefore a force of gravitation between the atoms which certainly pulls them together, just as a dropped ball is pulled towards the earth by gravitation. Is this force strong enough to hold the atoms together in the spoon? A child, or an adult for that matter, playing on the seashore knows how easy it is to start a mini-avalanche in a sandpile. The gravitational attraction between the grains of sand, which certainly is there, is not enough to maintain the shape of the pile when it is disturbed. One can conclude that the gravitational force between the atoms will not hold them together in the spoon. There must be some other force between the atoms that has nothing to do with their mass, and this force must lead to the rigidity of the solid. We shall see later that electrons and nuclei have a property called electric charge, or sometimes simply charge, and this is the source of the force that we are seeking.

In the past several decades, one of the exciting challenges in physics has been to understand how the properties of solids can be related to the way they are made up of atoms. This subject used to be, and sometimes still is, called *solid state physics.* A new name has come into use in recent years: *condensed matter physics.* The phrase 'condensed matter' means solids as well as liquids, both of which have atoms rather densely packed together. This similarity leads to some shared features of the way we can understand the properties of solids and of liquids. This branch of physics is not only one of the great intellectual achievements of the twentieth century; it has also led to practical applications that reach into many aspects of our everyday life. A famous example is the transistor. This is a device, made from a small piece of germanium or silicon, which began to replace valves (also called vacuum tubes) in radio receivers in the 1950s. A direct descendant of the transistor is today's chip, which has found its way into countless household items and tools of industry, and is at the heart of the modern computer. So pervasive is the chip now, that one can give the old saying a new meaning: 'Chips with everything!' Some other examples of practical applications which grew out of the understanding of the physics of condensed matter are: new magnetic materials used in electric motors, glass fibres carrying telephone conversations, liquid crystals and light-emitting diodes which display the numbers in pocket calculators, superconducting magnets for magnetic imaging systems which are used in medical diagnosis, high-strength alloys for jet engines.

I describe in the following chapters the atomic picture of solids which physics gives us. Some of you may have found that, if you ask a physicist-friend what she (or he, as the case may be) does, you are sometimes put off with the reply that you will not understand it if you do not know advanced mathematics. This attitude is, I think, not justifiable. It is true that a mastery of the subject in all of its intricate detail requires some heavy mathematics. I assume that you have no wish to acquire such mastery if you do not already have it. Rather, you would like to know the basic features of what physics tells us about the nature of solids. It is possible to describe these features using only words and pictures along with some arithmetic, and it is what I do in this book. After you have read it, you will have a picture of what a physicist sees in a piece of solid matter, and why it has the properties it does.

There will be no complicated mathematics in this book. I use only simple arithmetic, and sometimes symbols to denote numerical quantities, e.g. m to denote the mass (expressed in grams) of some object, x to denote a distance (in centimetres). Just as I add, subtract, multiply and divide with numbers, I do the same with symbols. I use the forms shown and explained below, for two numbers denoted by x and y:

$x+y$ means 'add x and y',
$x-y$ means 'subtract y from x',
$x\times y$ (or xy) means 'multiply x by y',
x/y (or $\frac{x}{y}$) means 'divide x by y'.

I can go one further step and denote the result of each of the above operations also by a symbol. For example, I could write the equation

$$x+y=z.$$

Here, if x equals 2 and y equals 3, then z equals 5. If x is 23.51 and y is 12.34, then z is 35.85. If I give any particular values to x and y, then the value of z is fixed. This single equation stands for every one of the separate operations of adding all possible pairs of numbers. The number of such operations is infinite, so that the equation is a very convenient shorthand indeed.

I give now another example of using such symbols, which illustrates what you will meet now and then in the book. An important property of a moving object is its speed. If the speed is constant, then I get the distance travelled by multiplying the speed by the time of travel. Using symbols, I write

$$x=v\times t$$

where x stands for the distance, v for the speed and t for the time of travel. This equation is read 'x equals v times t'. If the object happens to be moving at 10^6 centimetres per second, then I get the distance it travels in 2 seconds, say, by multiplying its speed, 10^6 cm/sec, by 2 sec. The equation applies equally well for any speed and any time of travel. You can see that the equation with its symbols gives us a convenient shorthand for the general statement, 'The distance travelled by any object which is moving at a constant speed is given by multiplying its speed by its time of travel.'

There is one other matter of notation that I have just used, when I wrote a number as 10^6, that may not be familiar to all of you. It is called the exponential (and when used in pocket calculators, the scientific) notation, and is useful in writing down numbers that are either very large or very small. For example, the number of atoms in one gram of a solid is about

$$1,000,000,000,000,000,000,000.$$

The width of an atom is about

$$\frac{1}{100,000,000}\text{ centimetre.}$$

The actual numbers will depend on what kind of atom it is, since the atoms of different elements are different in mass and size. In exponential notation, the two numbers above are written as follows:

$$1 \times 10^{21}$$

and

$$1 \times 10^{-8}\ \text{centimetre}$$

respectively and read as 'one times ten to the twenty-one' and 'one times ten to the minus eight centimetre' respectively. Two more examples: the number 123,000,000 is the same as 100,000,000 multiplied by 1.23, and in exponential notation is

$$1.23 \times 10^{8}$$

and the number 0.000456 is the same as 4.56 divided by 10,000, and in exponential notation is

$$4.56 \times 10^{-4}.$$

You will note that 10^8 is just a shorthand way of writing down the result of multiplying eight 10's together, and 10^{-4} is four (1/10)'s multiplied together. Other examples: x^4 means four x's multiplied together, and x^{-4} is four $(1/x)$'s multiplied together.

I now give you a kind of map for the journey we have just begun. In chapter II, I describe how the atoms are put together to form a solid. For this purpose I think of the atoms as tiny hard spheres, and do not worry about the nuclei and electrons. We shall find out in a later chapter how the atoms stick to one another to give a rigid solid, instead of sliding away into a heap. We shall understand why natural crystals like rock-salt and quartz have beautiful regular shapes with plane faces. It will turn out that solids can be divided into two types, called crystals and glasses respectively. The atoms in a crystalline solid find themselves in a highly ordered arrangement, somewhat like seedlings in a tree nursery (fig. I-1). The atoms in a glassy solid, on the other hand, are more like a crowd of people waiting to cross a road (fig. I-2); there is no noticeable

FIGURE I-1. The ordered arrangement of saplings in a tree nursery.

FIGURE I-2. A crowd of people waiting at a street crossing.

order in their positions. Most of the solids that we shall be dealing with in this book are crystalline. In a natural crystal of quartz, the regularity of the faces is due to the underlying orderly arrangement of its atoms. If you look at the surface of a silver spoon through a microscope, you will find that it is made up of tiny pieces of silver stuck together. I assume here that a metallurgist has treated the surface with some chemicals to show up the boundaries between these pieces. Now if you look at one of these pieces through an electron microscope which provides even greater magnification, you will discover that the silver atoms in this piece are sitting in an orderly arrangement: it is a crystal. We shall be considering crystalline solids in most of what follows.

I have so far talked only about atoms. There are around us not only objects made up of just one particular kind of atom – a copper wire, for example – but also objects made up of clusters of atoms which are called molecules. Quartz, for example, consists of molecules each of which has one atom of silicon and two of oxygen. We shall be talking mostly about atomic solids in this book, and you can assume, unless otherwise stated, that the picture will hold true for molecular solids too.

In chapter III, I introduce you to a set of ideas that are at the very heart of the physicist's understanding of nature. If I were to think of the journey through this book as a ramble on a mountain during which you will get beautiful views of the landscape, then this chapter will be the initial ascent. When you have finished the chapter, you will have a picture of what the physicist means when he talks about quantum mechanics. Just as English is the language of Shakespeare, quantum mechanics is the language of physics. I describe its basic ideas, using only words and pictures and some arithmetic. It is a way of looking at nature which is rather different from what we are used to in our everyday experience. It may therefore seem to you somewhat bizarre. However, everything we know about nature tells us that quantum mechanics is the correct description of it. When I say 'nature' here, I leave out living things. There are some scientists who say that since living things are composed of the same kinds of atoms as is inanimate matter, quantum mechanics should be able to explain their characteristics as completely as it does those of inanimate things. A very complicated matter, this, and the subject of much debate involving not only scientists but also philosophers and theologians. The question lies rather outside the scope of this book, and so I shall not go further into it.

In chapter IV, I apply the ideas of chapter III to an atom and find that quantum mechanics tells us that the atom is not at all what we had imagined it to be, namely a tiny hard ball. The philosopher Democritus said a long time ago that if one took a piece of matter and divided it into smaller and still smaller pieces, one would finally get to a piece that was so small that it could not be divided any further, and therefore he called it an atom. One might then think that an atom of, say, silver would be a tiny speck of silver with a silvery sheen, quite different from a copper atom which would be reddish in colour, or a glittering atom of gold. The reality turns out to be quite different, as we shall see in chapter IV.

A piece of solid weighing a gram has about 10^{21} atoms, which is a very large number indeed. There is no way in which we can hope to describe this piece in terms of what each atom in it is doing. Consider this: astrophysicists tell us that the big bang, which was the beginning of our universe, took place 10^{10} years ago. That is about 3×10^{17} seconds. So, even if a computer existed which took only 10^{-4} seconds to calculate and record the state of each atom, it would need $10^{21} \times 10^{-4} = 10^{17}$ seconds to complete the job – about a third of the life of the universe. In fact, there is no good reason for wanting such detailed information, other than perhaps the one given by a climber for wanting to climb Mount Everest: 'Because it is there.' Clearly we need some other way of dealing with such large numbers of atoms. Such a way exists, and you have already seen it applied to, say, the characteristics of the population of a country. Suppose I want to talk about the individual ages of the 248 million inhabitants of the United States. I do not even begin the hopeless task of listing the age of each person, all 248 million of them. Rather, I make a list of the numbers of people whose ages lie between specified limits, say 0–9 years, 10–19 years, and so on. I now have a compact description of the age distribution of the population, and have taken the first step into the subject called *statistics*. When we use this method in physics, we call it *statistical physics*. I shall introduce you to this topic in chapter V. In chapter VI, I replace the hard-sphere atom of chapter II with the quantum-mechanical atom of chapter IV, apply the statistical ideas of chapter V, and discover what happens to the electrons when the atoms find themselves in the solid. By doing so, I arrive at a picture of what a solid is like. It will look very different from the orderly stack of spherical atoms which I had in chapter II. I find that this new picture gives an explanation of why the solid holds together and does not disintegrate into a pile of atoms.

I compared earlier the passage through this book to a mountain ramble. When you reach the end of chapter VI, you will have reached an altitude with many views. You will be able to enjoy these views in the succeeding chapters, where the quantum-mechanical description that has been developed is used to explain the various properties of solids mentioned earlier. If we both complete the ramble successfully, you will be able to answer the questions with which this chapter opened, and many more. You will have a new way of looking at the world around you, a way which has its own beauty and which adds to the world-views of art, music, and literature.

You will find, as you read on, questions interspersed in the text. You can decide whether or not to try to answer each question first before looking at my answer which follows. In any event, please do not skip over the questions; they do form an integral part of the whole story.

The last section in each of the following chapters summarizes the chapter. You could read this section before beginning the chapter to get a general idea of what it contains, or read it at the end to help you put together what you have just read in the chapter. Alternatively, you could read it at the beginning as well as at the end to enjoy both benefits.

II CRYSTALS

1 A piece of silver metal

Mildred, a friend who is a physicist, wants to do some experiments with a crystal of silver. So she goes to her laboratory, and selects a container made of a ceramic which can stand being heated to very high temperatures. She puts some chunks of silver metal in it, and heats it up to a temperature of 1000 °C. She has to use a special thermometer for measuring such high temperatures, for an ordinary thermometer such as is used in the kitchen will not do the job. The silver melts into a liquid that collects at the bottom of the container (fig. II-1). The silver atoms are closely packed together, but not in fixed positions. They can slide easily past one another, which is why a liquid takes the shape of its container. She now lets the container cool down slowly, so that its temperature drops at a rate of less than 10 °C per hour. Then at a temperature of about 961 °C the puddle will freeze into a solid, as in fig. II-2(*a*). She waits until the piece of silver is cool enough to handle, takes it out and looks at it through a special microscope that can pick out the positions of single atoms as shown in fig. II-2(*b*). She finds that the atoms are sitting in an orderly arrangement: she now has a silver crystal. If the puddle was

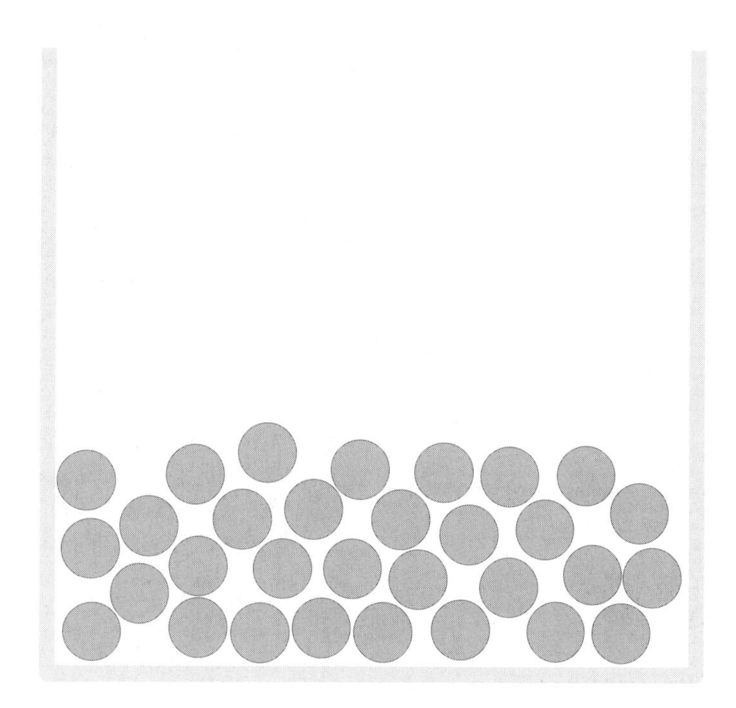

FIGURE II-1. The circles depict the atoms in a pool of molten silver.

FIGURE II-2(*a*). The circles depict the atoms in a slab of single-crystal silver. Note the orderly arrangement of the atoms, in contrast to the arrangement in fig. II-1.

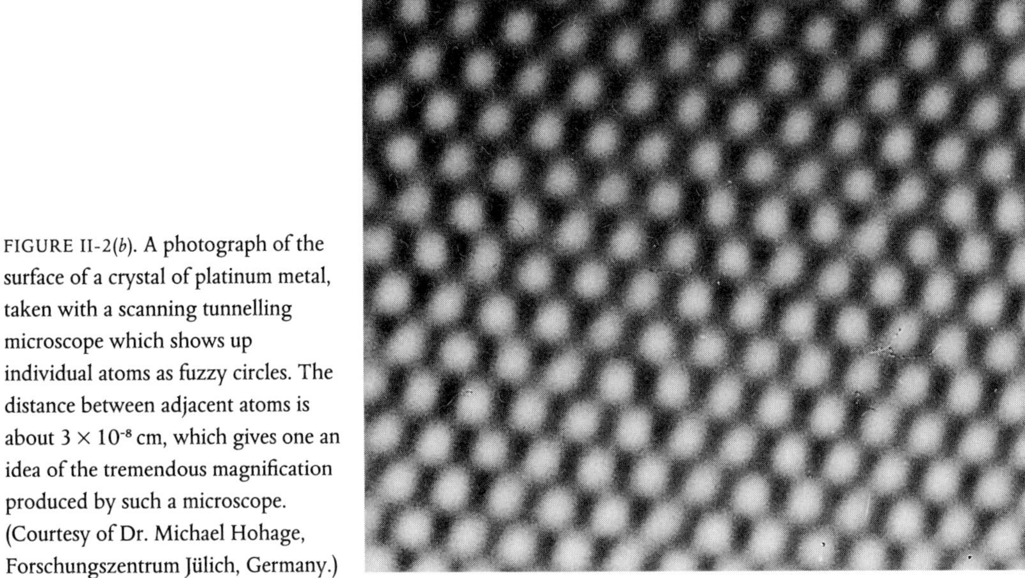

FIGURE II-2(*b*). A photograph of the surface of a crystal of platinum metal, taken with a scanning tunnelling microscope which shows up individual atoms as fuzzy circles. The distance between adjacent atoms is about 3×10^{-8} cm, which gives one an idea of the tremendous magnification produced by such a microscope. (Courtesy of Dr. Michael Hohage, Forschungszentrum Jülich, Germany.)

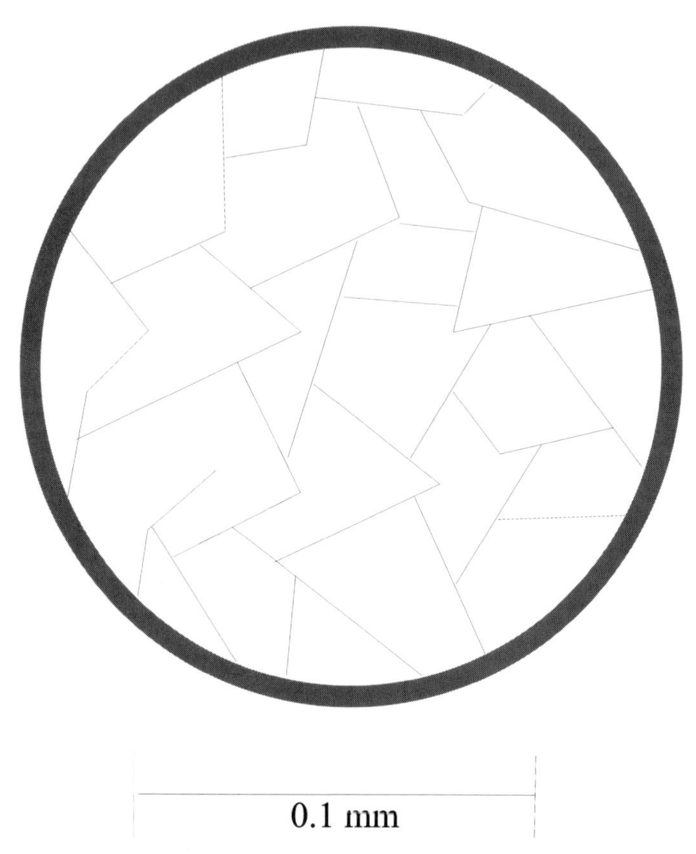

FIGURE II-3. Grains on the surface of a solid as they might appear through a microscope. The grains are shown as being a small fraction of a millimetre across, but the size in a given solid can vary depending on the details of how the solid is prepared.

shaped like a pancake, so is the piece of solid silver. It certainly does not look like a crystal in the everyday sense in which we use the term, namely a regularly shaped object with flat faces. It is nevertheless a crystal because of the orderly arrangement of its atoms, which is the essential feature of a crystal.

To end up with a piece of silver that is a single crystal, the rate of cooling has to be less than about 10 °C per hour. If the cooling is much faster, the solid will consist of many tiny pieces (which we call *grains*) stuck together, but each grain will itself be a crystal (fig. II-3). This is the form, which we call *polycrystalline* (composed of many crystals), in which everyday metal objects are.

Having come this far, Mildred is curious to find out what will happen if she cools the liquid even faster. So she builds a complicated gadget, with which she can cool the melted silver at a dizzying speed of something like a degree Celsius per 10^{-6} seconds. What is the result? A piece of silver in which the atoms are sitting higgledy-piggledy, with no noticeable order (fig. II-4). She has produced silver in the *glassy state*. It is so named because the molecules in ordinary glass are also in disordered positions. Glassy silver will still look *metallic*, and not like ordinary glass. Many of its properties will be

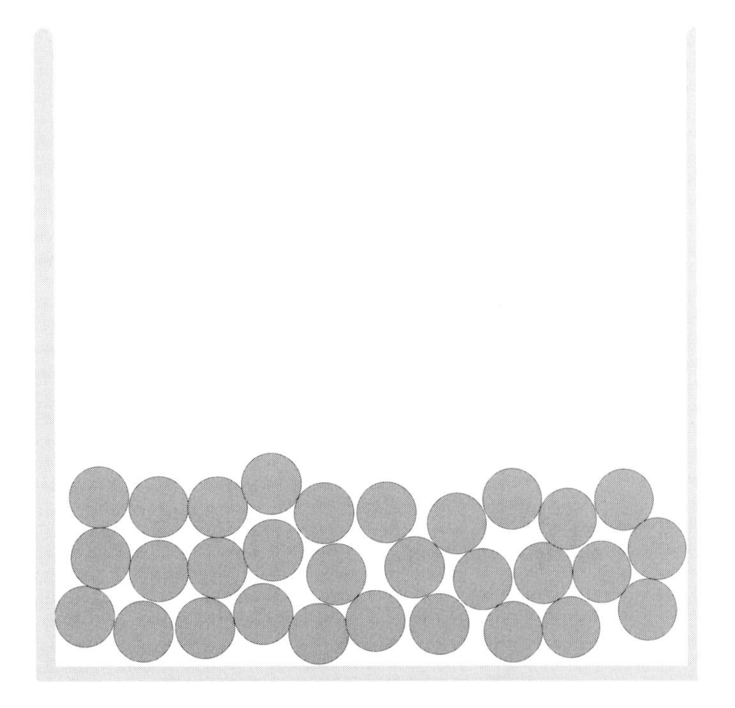

FIGURE II-4. The disorderly arrangement of atoms in a glassy solid.

found to be different from those of ordinary silver. This sort of cooling goes under the onomatopoeic name of 'splatt-cooling'.

Let us now try to understand how, depending on the rate of cooling, Mildred ends up with one of three forms of the solid: a single crystal, or a polycrystal, or a glassy substance. In doing so, we get an idea of how a physicist makes a picture of what happens in an experiment. We consider first the case of very slow cooling. As the cooling goes on, a very tiny crystal, consisting of a few tens of atoms, is formed at some point on the inside surface of the canister. With further cooling, this crystal grows by accreting to itself successive layers of atoms from the liquid, all sitting in the same orderly pattern with which the original tiny crystal began, until all the liquid is frozen. We thus end up with a single crystal.

We now come to the case of intermediate cooling rates. Things are happening faster now, and in an actual experiment there will be unavoidable fluctuations and disturbances. The freezing temperature will be reached first not at just one point as in the last case, but at a large number of points on the inside surface of the container as well as in the liquid at about the same time. Tiny crystals will start growing at each of these points and finally all the liquid is frozen into a polycrystal. Each grain is a single crystal with its atoms in the same orderly arrangement, but the direction of this order will change from grain to grain.

I illustrate what is meant by 'the direction of the order' with an analogy, a troop of one thousand soldiers on a parade ground. Suppose I ask them to fall into parade formation. They would then form a single orderly pattern of rows and columns. If on the other hand I divide them into ten units, each of one hundred soldiers, and then ask them to fall into parade formation at ten different points on the parade ground, but do not tell them in which direction they should face, I would end up with a different result. It is true that each unit will look exactly like every other unit, but while all the soldiers in a given unit will be facing in the same direction, this direction is very likely to be different for the different units. Something like this is what happens when the liquid is cooled rapidly, leading to a polycrystalline solid. It is possible to assign an 'orientation' to each grain in such a solid, somewhat like calling the direction in which the soldiers of a given unit are facing as the 'orientation' of the unit. If we look at the ordered pattern of atoms in the whole crystal, we find that the order will be different in different directions, analogous to what is illustrated in fig. I-1. It shows a tree nursery, with the seedlings planted in a regular pattern. You can see that the plants are spaced apart by different distances in different directions, so that it is possible to distinguish one direction from another.

Finally we come to the splatt-cooled solid. The atoms in the liquid have no order in their positions at any given instant of time. The positions however are changing continuously with time as the atoms slide past one another. When splatt-cooling takes place, the temperature drops so fast that the atoms are stopped dead in the last positions they had in the liquid, which were not in any orderly arrangement. The result is a glassy solid.

I can also look in another way at what the difference is which causes the liquid to end up as a crystal or as a glass: namely in terms of what the atoms are doing as time goes on in the two cases. If I do the freezing slowly enough, the atoms have enough time to find their way to the positions in the solid where they would most like to be, so to speak, and the result is a crystal. If on the other hand I speed up the process sufficiently, the atoms just do not have enough time to find their way to their preferred crystalline arrangement. So they freeze into the disordered arrangement of a glass. It is known from actual experience that, given a long enough time, the atoms in the glassy solid will slowly rearrange themselves eventually into an ordered crystalline arrangement. One possible crystalline form of ordinary glass is quartz. The glass in some archaeological objects, known to be many centuries old, is seen to have transformed to quartz, which is the crystalline form of glass. This process is known as devitrification. It appears therefore that the solid, given enough time, always ends up being crystalline.

So now the question arises, why do the atoms prefer to be in an ordered crystalline arrangement? We shall find the answer in the next section.

2 The crystalline arrangement of atoms

I have said that when the liquid is cooled slowly to form a solid, the atoms end up in their preferred positions, which is the orderly arrangement of a crystal. There are two features which characterise this arrangement.

The first of these follows from the fact that the solid is stable: it does not spontaneously change its shape. This means that the atoms continue to stay where they are with respect to one another, and do not move away from these positions: the net force on each atom is zero. If I take a rod and pull on its two ends, it will get a bit longer; if I push, it will get shorter. The actual change in size is perhaps invisible to the naked eye, but it has been measured with special instruments. In either case, the rod returns to its original shape when I let go. If I describe this behaviour in terms of what the atoms are doing, I would say that when the rod is stretched the atoms move a bit further apart, and a bit closer together when it is compressed. An ordinary coil spring becomes shorter when compressed, longer when stretched, and in each case returns to its original length when released. I can explain the behaviour of the rod by imagining that there are springs connecting the atoms to one another. There are of course no real springs in the solid; it is just that the influence of the atoms on one another due to being made up of charged nuclei and electrons produces the same effect, and we might talk about virtual springs connecting the atoms to one another. This property of a solid, of recovering its shape after being stretched or compressed, is called its *elasticity*.

QUESTION But what about a copper wire? When I bend it, it remains bent and does not return to its original shape. This must surely mean that at least some of the atoms have moved from their original positions.

ANSWER The reason the wire remains bent is that the grains in the wire are not perfect crystals, but have small imperfections which permit some atoms to move about when the wire is bent. We shall learn about these imperfections in chapter VII, and see how they can explain the bending of a copper wire. These imperfections have very little effect on many of the properties of solids that we shall be considering.

The second feature of the crystal comes from the fact that all atoms of silver are identical. More generally, all atoms of any element are identical, but different from atoms of any other element. Now suppose that I place myself in the position of a particular atom in the crystal, and look at the atoms around me. They will be distributed in their preferred pattern at various distances in various directions around me. Now I move to another atom in the crystal. Since all the atoms are identical, the preferred pattern of atoms around this second atom should be exactly the same as around the first atom. Whatever factors led to the preferred arrangement around the first atom should hold equally for the second atom and lead to the same arrangement. The crystal, on an atomic scale, must look exactly the same from the viewpoint of every atom in it. The universe around us seems to have a similar property. When considered on a large scale that includes enough stars, galaxies and so on, the universe looks the same from one point in it as from any other point. Astronomers call this the *cosmological principle*. The atoms in a crystal satisfy their own version of the cosmological principle. There is nothing mystical about this principle as it applies to solids. It is merely a consequence of the experimentally established fact that all atoms of the same element are identical.

So we now have the two general characteristics of a crystal: the atoms in it are as if connected to one another by virtual springs, and their positions satisfy the cosmological principle. When we put atoms together to meet these two requirements, we shall end up with the ordered atomic arrangement in a crystal. We shall build the crystal in three stages, for this makes it easy to satisfy the two requirements. We shall first build a straight line of identical atoms, then assemble a number of identical lines of atoms to give a plane of atoms. Finally we shall stack identical planes of atoms above one another to give a crystal, always making sure at each stage that the two requirements are met.

3 The one-dimensional crystal

I now introduce an object that is very useful when trying to understand nature, even though it does not exist in nature. This is a *one-dimensional crystal*, consisting of a long row of atoms sitting equally spaced in a straight line. A straight line has only length, and zero width and zero thickness. Suppose that I measure distances along the line from one of its ends. Then to tell you where any point on the line is, I need to give you only its distance from that end. That means that just one number, namely this distance, is enough to tell you exactly where that point is. A 'point' in this context is a thing that has a position in space, but has vanishingly small length, width and thickness. Since we need only one number to tell us where a particular point is on a straight line, we say that it is in *one-dimensional space.* We assume that the atom is a point with a mass equal to the mass of the atom, and that the influence of a neighbouring atom on a given atom is as if a spring connects the two.

I show in fig. II-5 three stages labelled A, B, and C in building up a one-dimensional crystal. In stage A, I set down atom 1 and put atom 2 on its right at a distance a. Since

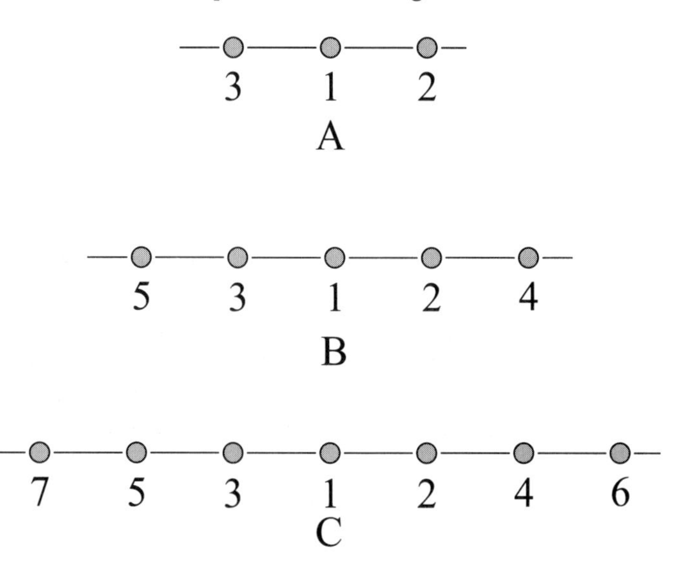

FIGURE II-5. Three successive stages, labelled A, B, and C, in building up a one-dimensional lattice of atoms. The distance between adjacent atoms is a.

atom 2 now has an atom on its left, so must atom 1 and at the same distance. So I put down atom 3 to the left of 1 at a distance a from it. Now atom 1 has an atom on each side, and therefore so must atoms 2 and 3, and this is satisfied by atoms 4 and 5 in stage B. Continued application of the cosmological principle will lead to stage C and so on, so that we get a long straight line of equally spaced atoms, with neighbouring atoms connected by virtual springs of length a and no net force on each atom.

We can now deduce some features of the virtual springs from the properties of this one-dimensional crystal. We assume that it has its undisturbed length, i.e. it can be stretched or compressed, but it will return to its original length afterwards. This clearly means that the undisturbed length of each spring is a, the spacing between the atoms. Since the virtual springs are just a way of expressing the influence of one atom on another, and as the atoms are identical, all the springs must also be identical in their strength. I mean by the strength of a spring how hard it is to press or to stretch it.

QUESTION Will the length of the springs be the same for different kinds of atoms, say silver and copper?

ANSWER No. The length and the strength of the spring will be determined by the detailed structure (in terms of nuclei and electrons) of the two atoms it connects, because this is what determines how one influences the other.

QUESTION The lattice must have two ends somewhere, and the atoms near there will certainly not satisfy the cosmological principle. Is this not a problem?

ANSWER The influence of one atom on another weakens rapidly as their distance apart increases, so that a given atom can feel the effect of only atoms that are a few times the distance a from it, and is blissfully unaware of atoms further away. So it suffices if the cosmological principle applies for a few neighbours surrounding each atom. The condition will then hold for all the atoms except a few at the ends. We shall assume that the total number of atoms is very large compared to the small number at the ends, so that we can in practice say that all the atoms satisfy the cosmological principle.

In any crystal, there are some atoms at its boundaries. If the crystal is one-dimensional, they are at its two ends. They are at the edges of a two-dimensional crystal, and on the surfaces of a three-dimensional crystal. We shall assume in everything that follows that we are looking at the properties in a region of a crystal far enough away from its boundaries so that effects due to the atoms at those boundaries can be neglected. Such properties are called *bulk properties*, to distinguish them from *surface properties* which are determined by the atoms at the boundaries.

The distance a between adjacent atoms is called the *lattice spacing* of the crystal. I have introduced here a new term, *lattice*. When I am interested only in the relative positions of the atoms, but not in a specific value of a or what kind of atoms they are, I call it a lattice.

4 Two-dimensional lattices

The one-dimensional lattice is a suitable building-block with which to assemble a two-dimensional lattice, as it already satisfies the cosmological principle and has the virtual springs built into it. So I make myself a large number of identical one-dimensional lattices, each with spacing *a*. I label these lattices 1, 2, 3, and so on. I want to assemble these on a flat table-top to give me an object that has both length and breadth, and in such a way that the cosmological condition continues to be satisfied. I begin by setting down lattice 1 on the plane. Then I place lattices 2 and 3 on either side of lattice 1, parallel to and at a distance *b* from it (fig. II-6, shaded portion). I imagine that corresponding atoms between the lattices are connected by virtual springs of length *b*. Further, I make each of the two angles that I have marked on the diagram equal to 90 degrees. We shall see later that these angles can also be different from 90 degrees, and then one would get a different lattice.

QUESTION Why is the distance *b* taken to be different from the distance *a*? All the atoms are identical and so one might expect the virtual springs to have the same length.

ANSWER Remember that the virtual springs stand for the mutual effect of the electrons and nuclei of the neighbouring atoms on one another. The electrons are distributed differently in different directions around the

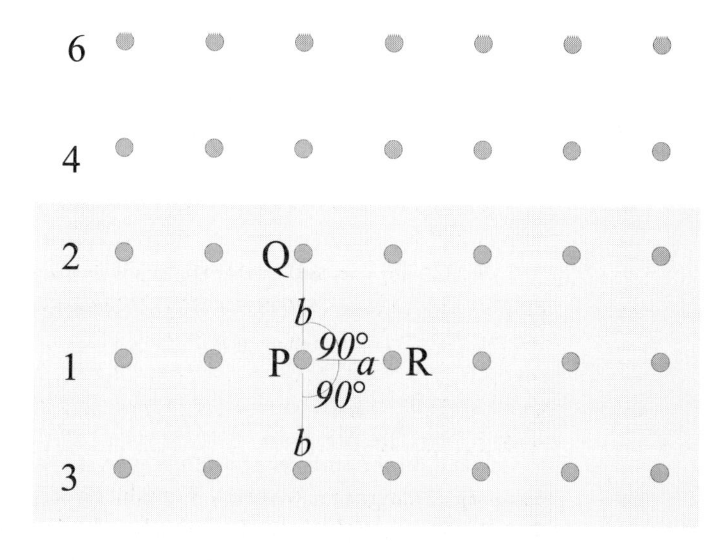

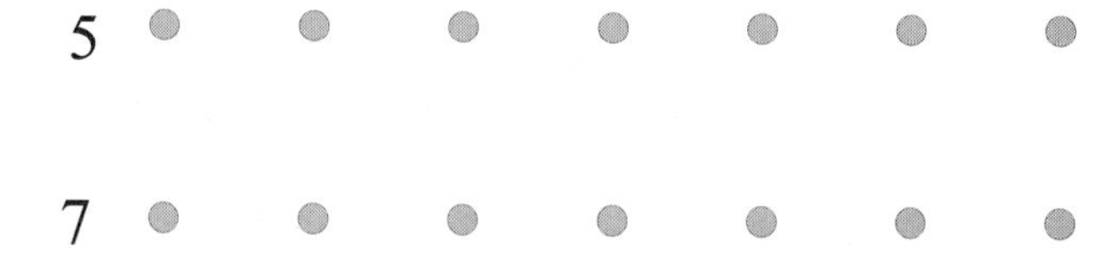

FIGURE II-6. The stages in building up a two-dimensional lattice, beginning with the three one-dimensional lattices labelled 1, 2 and 3 (shaded portion).

nucleus. This can result in the virtual spring between P and Q having a different length and strength from the one between P and R. It is possible as a special case to have *a* and *b* be the same, as we shall see later.

You can see in the shaded part of fig. II-6 that each atom in lattice 1 has eight near neighbours around it, arranged in exactly the same pattern. Thus the cosmological principle holds for the atoms in lattice 1, but not for the atoms in lattices 2 and 3 which have only five nearest neighbours each. However, we know how to solve this problem, from our experience in building the one-dimensional lattice. We add pairs of lattices 4 and 5, 6 and 7 (the rest of fig. II-6), and so on. The arrangement of atoms around each atom is now exactly the same, and there is no net force on it; we have constructed a *two-dimensional crystal.* We can completely specify the crystal by giving the kind of atom (silver, copper, etc.), the lengths *a* and *b*, and the angle of 90 degrees. We call *a* and *b* the *lattice parameters.* The numbers, of atoms in each row and of rows, only specify the size of the lattice, and have nothing to do with how the atoms are arranged.

We could now go on to the next stage, and complete the construction of a solid crystal by stacking a large number of two-dimensional crystals on top of one another in such a way that each atom still feels no force from its neighbours. Before doing that, I should like us to take a closer look at the two-dimensional crystal that we have before us. We shall see that such a crystal has some beautiful and important features which carry over, *mutatis mutandis,* into the solid crystal. To do so, we shall be using pictures. We can make an exact picture of the two-dimensional crystal on a page, and see those features directly. It is not so easy to do so with a picture of a solid crystal: that is, unless your imagination can leap easily from the flat page into the third dimension.

We use the term *lattice* when we are interested only in the relative positions of the atoms, but not in their kind. So the lattice is specified by the lengths *a* and *b* being unequal, and the angle of 90 degrees. What we have just finished constructing is a *rectangular lattice.* The whole lattice can be broken up into a lot of rectangles, each with edges *a* and *b.*

Is the rectangular lattice the only one I can have on a plane, just as there was only one lattice I could construct on a straight line? No, because I can change the angle and the ratio of *a* to *b,* which determine the lattice. I denote the angle by the Greek letter α (alpha). I can now produce four more lattices, to give me a total of five. Each of them shows order, but in a way which is different from the others. I list below the five lattices with their names.

1. *Simple rectangular lattice.* *a* and *b* are different, α is 90 degrees.
2. *Centred rectangular lattice.* I get this by adding an additional atom at the exact centre of each rectangle formed by four atoms from adjacent rows of the rectangular lattice.
3. *Square lattice.* *a* and *b* are equal, and I use the same symbol *a* for both of them. The angle α is again 90 degrees.
4. *Triangular lattice.* *a* and *b* are again equal, but α is now 60 degrees.

5. *Oblique lattice.* a and b are unequal, and α is some angle which is different from 90 degrees.

Figure II-7 shows the positions of the atoms in the basic unit for each of these lattices. This unit is called the *unit cell* of the corresponding lattice. It is the smallest part of the

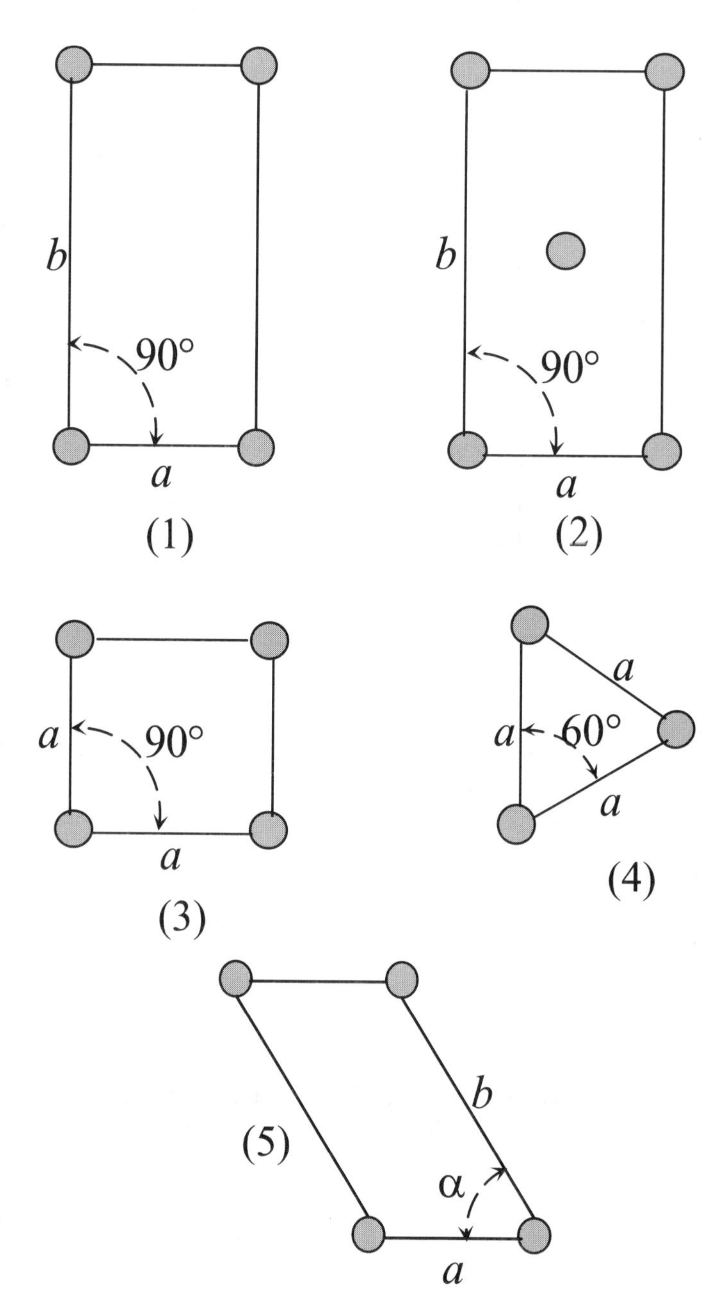

FIGURE II-7. Unit cells of the five two-dimensional lattices: (1) simple rectangular; (2) centred rectangular; (3) square; (4) triangular; and (5) oblique.

lattice which completely specifies what the whole lattice looks like. I can build up the lattice by assembling the unit cells rather like tiles on a floor. Each atom at the edge or the corner of a unit cell is then shared with the immediately adjacent unit cell or cells.

QUESTION Why are there no more two-dimensional lattices?
ANSWER These five are the only ones possible which satisfy the cosmological principle.

What distinguishes each of these lattices is its *symmetry*, a term that I shall explain in the next section. If I try to make another lattice that seems to look different from any of these, it will turn out that it is just one of these, but with changed values of the lattice parameters and sometimes the angle. For example, if I add an atom at the centre of each little square in the square lattice, I again get a square lattice, but with a different lattice parameter, as you can verify by drawing a picture.

QUESTION What if the added atom is not at the centre of each square?
ANSWER The arrangement of atoms that results does not satisfy the cosmological principle, because the surroundings of each added atom are different from those of each atom in the original lattice.

You may amuse yourself by seeing what happens if you add an atom to the centre of each little triangle in the triangular lattice. By drawing a picture, you will find that the result is again a triangular lattice, but with a smaller lattice parameter.

5 Order and symmetry

I have been talking about the *order* in a crystal or a lattice, implicitly appealing to your innate sense of what the term means. However, you might well say that order lies in the eyes of the beholder. To take an example from the world of hearing rather than of sight, there are those who find the music of Johann Sebastian Bach highly ordered, and that of John Cage somewhat less so. But there are others who detect much order in Cage's music too. Such differences of perception contribute to the richness of the world of art, but could lead to undesirable confusion in the world of physics. We need to make more precise what we mean by the order in a crystal. We do so by introducing the idea of *symmetry*, an idea which plays a central role in physics.

In our everyday life, we think of some objects as having a certain symmetry to them, and others not. For example, suppose I look at two objects on a table, a simple vase and an irregular lump of clay as in fig. II-8, part (*a*). I shall get to the vase under the table later. I say that the vase on the table has symmetry, but the lump has none. The following is what I mean by this: suppose I look away from the objects, and you rotate both the vase and the lump without moving them from their places on the table. The result is shown in fig. II-8, part (*b*). When I look at them again, the lump will look different, but the vase will look exactly the same as before. I assume here that the vase is absolutely

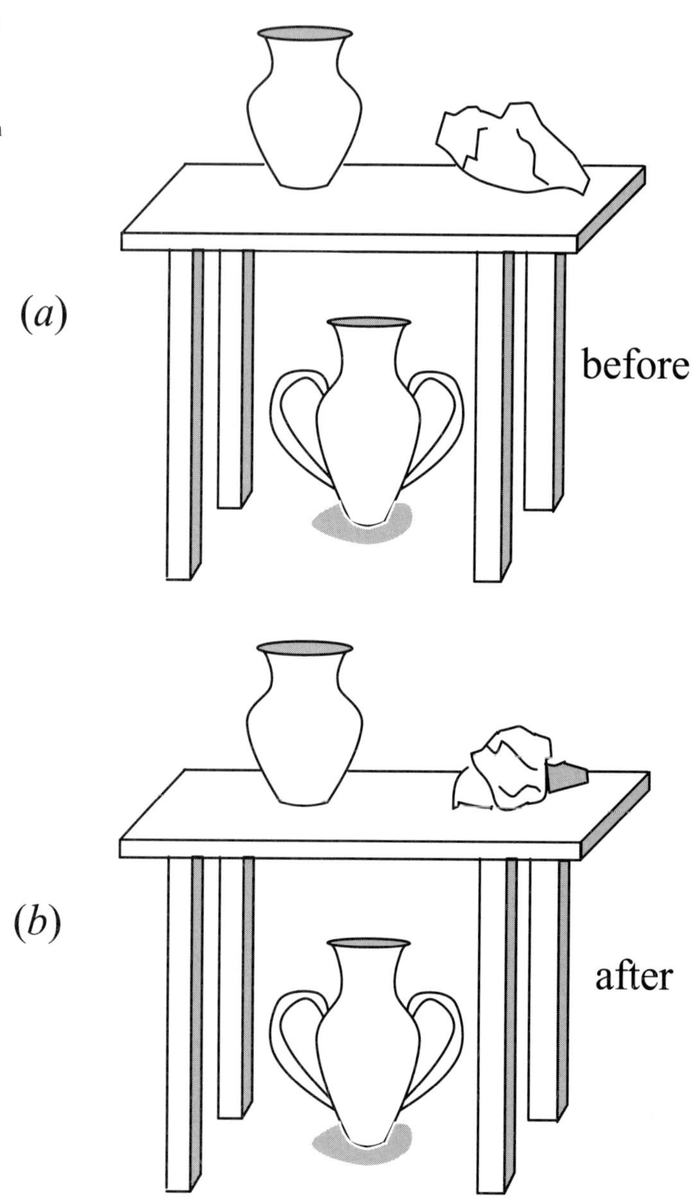

FIGURE II-8. To illustrate the idea of symmetry. Part (*a*) is before, and part (*b*) is after, the vases and the lump of clay have been rotated as described in the text. The vases have symmetry, the lump has none, under these operations.

perfect, with no distinguishing marks or patterns on it, and that the lump is – well, a lump. There is only one axis of rotation for the vase which will lead to the above result. If you had rotated it by turning it upside down, it would look different when I look at it again. I summarize all this in the following statement:

The vase has symmetry for rotation by any angle about one particular axis, and the lump does not.

I can explain the term *axis of rotation* by an example: the axis of rotation of the earth is the straight line joining its north and south poles.

I consider another example: a one-dimensional lattice. Look at a section of the lattice far from its ends, and imagine that the entire lattice is moved a distance along its length equal to the lattice spacing a, so that each point is now in the position where its neighbour was before. The lattice looks exactly as it was before. I express this property of the lattice as follows:

A one-dimensional lattice has symmetry for a translation of one lattice spacing along its length.

I mean by the term *translation* the moving of a body from here to there without rotating it.

As a final example, consider the vase under the table, with a pair of identical handles on opposite sides (fig. II-8). This vase has an axis of rotational symmetry only for half of a full rotation, namely through an angle of 180 degrees. A rotation through a further 180 degrees brings the vase back to its starting position. Thus there are two positions, each reached from the other by the same angle of rotation, in which the vase looks exactly the same. I describe this property as follows:

A vase with a pair of identical handles on opposite sides has a two-fold axis of rotational symmetry.

QUESTION Can you see a plane of reflection symmetry in this vase?

ANSWER Yes, a plane passing through the axis and perpendicular to the plane joining the two handles.

These examples suggest a precise way in which we can talk about the symmetry of a given object. First, we perform a specific *operation* on it. This operation can be a *translation*, a *rotation* or a *reflection*. If at the end of the operation the object looks exactly the way it did before, then we say that the object is symmetric under that specific operation. This is the sense in which we shall talk about the symmetries of lattices and crystals. When we say that the crystal has order, we mean that it has certain symmetries.

6 Symmetries in a two-dimensional lattice

We now look at the symmetries of the lattice, a portion of which is shown in part (a) of fig. II-9. Imagine identical atoms sitting at each lattice point. We shall first consider its translational symmetry. I move the lattice to the right, such that the atom marked A finds itself in the position where the atom marked B was, and B moves to where C was, and so on. The crystal now looks as it did before the translation. We have thus found a translational symmetry of the lattice, but it is only one of many. For example, if I make a translation which moves atom A to the position of atom P, then each other atom moves to the position of another atom which was in the same relative position to it as P was to A. This is again a translational symmetry, but different from the first one, because the size of the displacement and its direction are different. A third translational

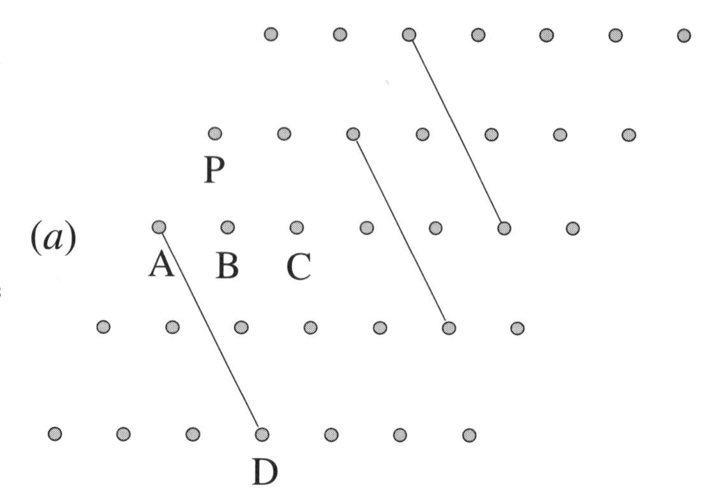

FIGURE II-9. A two-dimensional lattice. Part (*a*) illustrates its symmetry under translations which take A to B, or to P, or to D. How two other translations look when A goes to D are shown by lines connecting the relevant lattice points. Part (*b*) shows how the same lattice can be constructed from different sets of lines of atoms, as indicated by the dotted lines and dashed lines. The directions AB and AC are close-packed, and AD is not.

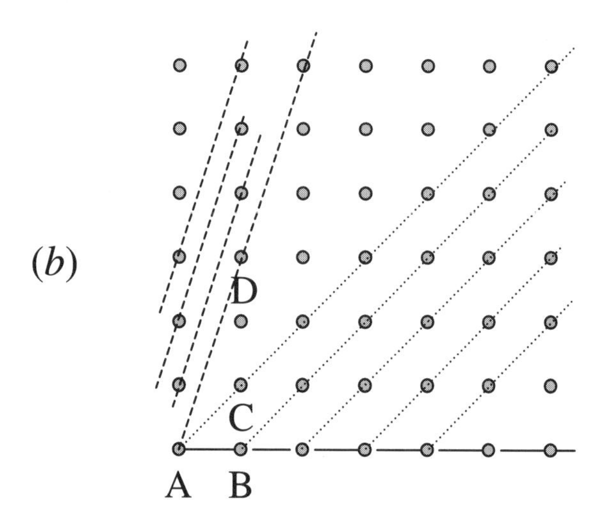

symmetry is one that takes atom A to the position of atom D. In fact, any translation that takes a given atom to the position of *any other atom* is a symmetry operation for the lattice. We can summarize:

Two-dimensional lattices have translational symmetry for any translation that takes a lattice point to the position of any other lattice point.

Closely related to this translational symmetry is a property of a lattice that gives us a hint why naturally occurring crystals have regular *crystalline* shapes. Consider the lattice shown in part (*b*) of fig. II-9. If I join the atoms marked A and B by a straight line, and extend it in both directions, I get a row of atoms that are equally spaced. If

I do the same thing with atoms A and C, I again get a straight row, but with a larger spacing. With A and D, I get a straight row with still larger spacing. In fact, I get a straight row of equally spaced atoms if I start with any pair of atoms, whatever their distance apart, and carry out the above operation. Each of these rows is a one-dimensional lattice with its own characteristic spacing and direction. When the spacing is equal to, or only somewhat larger than, the lattice spacing, we talk about a *close-packed direction* in the lattice. These are very few in number. In the present case, AB and AC are certainly close-packed directions, AD is not. Notice that the lattice can be thought of as made up of a large number of any one of these different one-dimensional lattices. I have drawn some lines in the figure to help you see this.

The existence of a few close-packed directions leads to the shapes that natural crystals have. Such crystals are formed very slowly from the liquid. They have a small number of faces, of which some are in pairs parallel to each other. Even though two-dimensional crystals do not exist in nature, we might by analogy think that, if they did, they would have a few straight edges, with perhaps some in parallel pairs. Suppose we slowly freeze a two-dimensional liquid to form a crystal. At first a very tiny crystal, with just a few atoms, is formed. It proceeds to grow by adding atoms from the liquid. The average spacing of the atoms in the liquid is about the same as in the crystal. So the atoms from the liquid that join the crystal in successive layers are much more likely to find themselves in the few close-packed directions where their spacing is about the same as in the liquid. If a kink should develop in a row of atoms at a growing edge, further atoms will join the crystal there in positions that smooth out the kink rather than make it grow. This is because an added atom sees more neighbouring atoms near it at the kink than elsewhere, and is therefore more strongly attracted towards it. Figure II-10 shows photographs of stages in the growth of very thin crystals of platinum. One sees the straight edges that correspond to the close-packed directions.

After this detour into the shape of crystals, we return to the symmetries of the square lattice. I look at part (*a*) of fig. II-11 which shows a square lattice, and imagine an axis coming straight out of the paper and passing through the lattice point A. If I rotate the lattice through an angle of 90 degrees, which is one quarter of a full turn, about this axis, it looks exactly the same as before. I rotate it twice more, each time

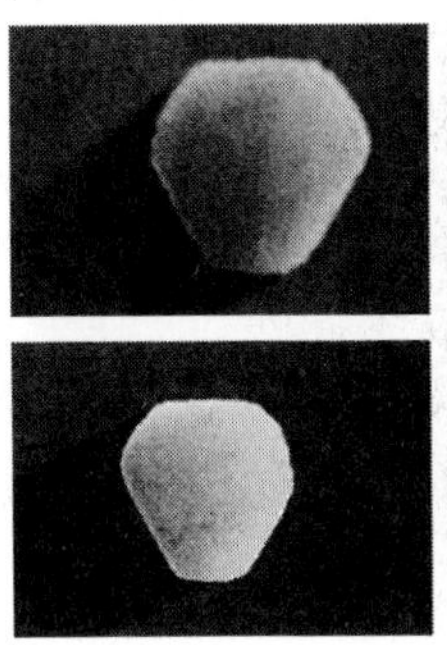

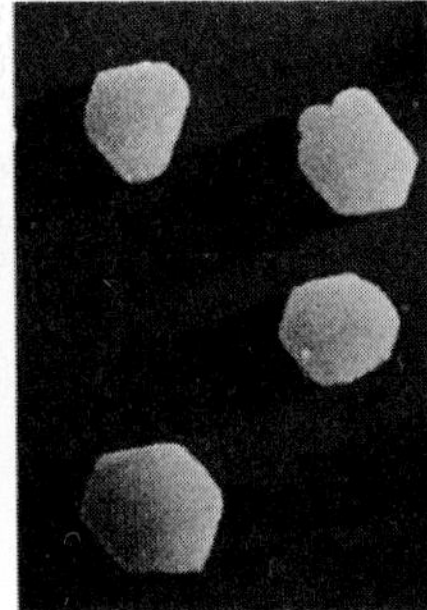

FIGURE II-10. Four pictures of small thin crystals of platinum metal formed by depositing atoms on a platinum surface. The edges are along close-packed directions in the lattice, and are about a hundred angstroms in length. The pictures were taken with a scanning tunnelling microscope. (Courtesy of Dr. Michael Hohage, Forschungszentrum Jülich, Germany.)

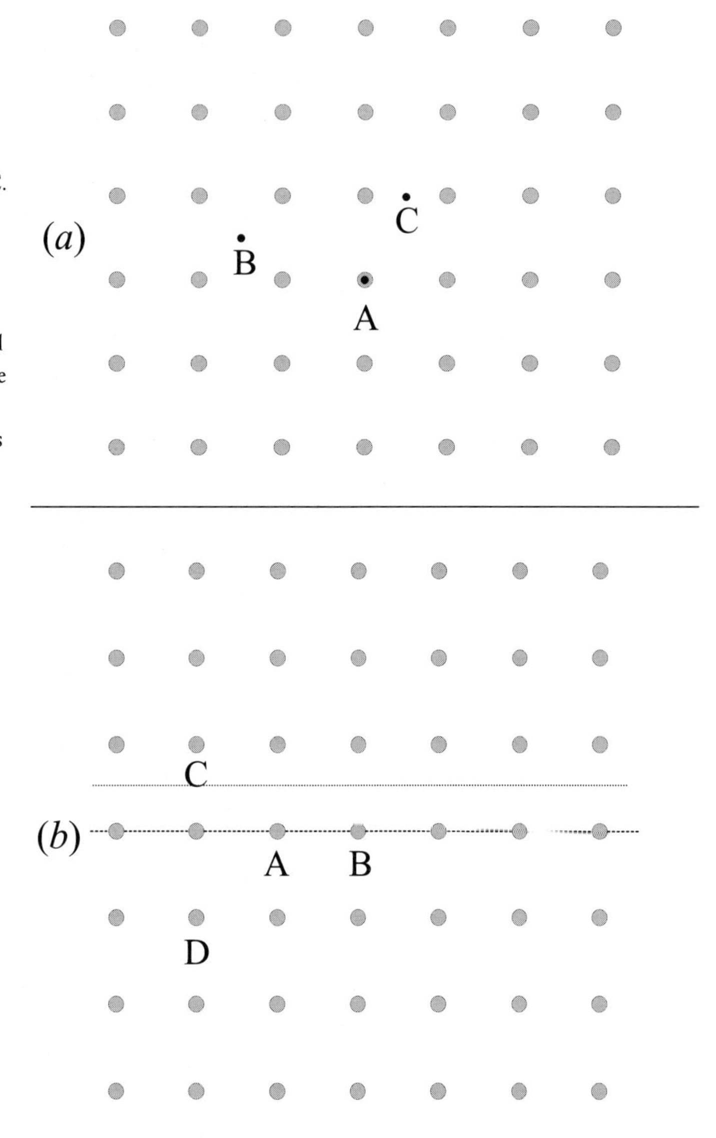

FIGURE II-11. Rotational and reflection symmetries in a two-dimensional square lattice. Part (*a*) shows axes of four-fold symmetry through the points A and B, and an axis of two-fold symmetry through C. An axis of *n*-fold symmetry means that the lattice can be rotated to *n* positions about that axis, and still look the same. In part (*b*) of the figure, the broken line through A and B is a line of reflection symmetry. The reflection interchanges atoms C and D, and all other similar pairs of atoms likewise, leaving the lattice looking the same as before. The dotted line above A is also a line of reflection symmetry.

through a further 90 degrees, and at the end of each rotation it still looks the same as before. A final rotation of 90 degrees brings it back to the starting position, because a 360-degree-rotation is a full rotation. I have thus found a rotational axis of four-fold symmetry: there are four positions reached by successive rotations through 90 degrees where the lattice looks the same. This axis of symmetry may pass through any lattice point.

QUESTION Are there any other axes of rotational symmetry for the square lattice?

ANSWER You can verify from the upper picture that if the axis passes through the centre of a unit cell, B, it is again an axis of four-fold symmetry. If it passes through the midpoint between two nearest neighbouring points, C, then it is an axis of two-fold symmetry.

QUESTION What are the rotational symmetries in the triangular lattice for an axis through a lattice point, and through the centre of a triangular unit cell?

ANSWER Six-fold and three-fold respectively.

These lattices have one other type of symmetry. Consider the lattice shown in part (*b*) of fig. II-11. Draw a straight line through A and B, think of it as a mirror, and imagine that the portions of the lattice lying above and below the mirror are reflected by it. Then, for example, the reflection of atom C will coincide with atom D, and vice versa. The reflection of the portion of the lattice that is above the mirror will look exactly like the portion below, and vice versa. The appearance of the lattice is unchanged by this reflection. The line through AB is a *line of reflection symmetry*.

QUESTION Can you find another line of reflection symmetry in the square lattice?

ANSWER Yes. The straight line through A and C.

QUESTION There are six lines of reflection symmetry in this lattice. Can you find the remaining four?

ANSWER A line parallel to the line joining A and B, and sitting midway between B and C. Now if you imagine this and the other two lines each rotated through an angle of 90 degrees, the resulting lines are also lines of reflection symmetry.

QUESTION Which of the five lattices we have been looking at has the least symmetry?

ANSWER The oblique lattice. It has only translational symmetry and one axis of two-fold rotational symmetry.

Let us pause and take stock of what we have done. I list the highlights:

1. The lattice surrounding a lattice site looks exactly the same as the lattice surrounding any other lattice site. This is the cosmological principle as applied to a crystal, and is a consequence of the fact that all the atoms of a given element are identical.
2. The following is what is meant by the *symmetry* of a lattice: I carry out a specified operation on the crystal. The operation may be a translation, or a rotation, or a reflection. If it is a symmetry operation, then the crystal will look exactly the same after the operation as before it.
3. The order in a lattice or crystal is the total of its symmetries.

4. Each of the two-dimensional lattices has its own characteristic set of translation, rotation and reflection symmetries.

I want now to say something about a more general matter. There are some physicists who, when they have the time, think and even talk or write about the beauty of physics. You might recall the physics you were exposed to in your school days, with its balls rolling down slopes, electric currents in wires causing nearby compass needles to move, and so on. You might wonder what exactly is supposed to be beautiful about all those things.

I think that there is beauty in the way symmetries suddenly appear when I put atoms together to form a solid. We began with a simple and very reasonable condition: each atom in the crystal should see the same pattern of atoms around it. Out of this emerged the variety of symmetries we see in the crystal. One ordinarily associates beauty with creations in the arts: a Mozart symphony, a Thyagaraja song, a Leonardo da Vinci painting, the Taj Mahal. If you can begin to think of the symmetries of a crystal also as being beautiful, you will have enlarged and enriched your sense of beauty.

There are other symmetries that exist in nature, in addition to the ones we have seen in crystals. Consider, for example, a container filled with helium gas. Each helium atom looks exactly like every other helium atom. Nothing about the gas would change if I imagined interchanging the positions of any two atoms in the gas. The gas is *symmetric* for such an operation. We shall see later that quantum mechanics gives this particular symmetry an unexpected new twist, with far-reaching consequences for our understanding of nature.

The following are two more examples of symmetries in nature. All objects exist in space and time. Their behaviour is governed by the laws of physics, such as the law of gravitation and the law of force between electric charges. We know from experience that these laws are exactly the same on earth as anywhere else in the universe, and have been the same for as far back in time as we can determine. The astronauts' experiments on the moon obeyed the same laws of physics as the experiments we do on earth, and the weights which Galileo dropped from the leaning tower of Pisa obeyed exactly the same law of gravitation as the pencil I just dropped on the floor. The laws of physics are the same at all points in space, and at all instants of time. From the way we have understood symmetry in physics, we can say:

The laws of physics are symmetric for displacements in space and in time.

Having got this far, you might be willing to grant that there is some sort of beauty in these symmetries, but you might well wonder what on earth one can do with them. These last symmetries have profound consequences for all of physics. For example, the symmetry with respect to time implies that energy cannot be created or destroyed but can only be changed from one form to another. To explore these matters fully would take us far afield from our present concern, which is to understand solids. So let me give an example of how we might use the symmetry of a crystal to simplify the task of understanding its properties. We saw that the neighbourhood of any one atom is exactly the same as that of every other atom in a crystal of any given element. We can use this symmetry to bypass the hopeless task of trying to describe what is happening around each of the about 10^{21} atoms in the piece. It should suffice to arrive at a description of the state of affairs around one atom, because the crystal symmetry then tells us that it should be exactly the same around all the others. The phrase 'state of affairs' is rather vague. Its meaning will become clear when we later get to talking about the structure of the crystal in terms of its electrons and nuclei.

7 Three-dimensional lattices

After this detour into the subject of symmetry, we continue with the next part of our task of assembling a crystal. We were left at the end of section 4 above with five two-dimensional lattices. I take one of them, say the rectangular lattice with lattice parameters *a* and *b*, and make a large number of identical copies. I stack them now on top of one another and end up with a three-dimensional lattice (with length, breadth as well as height), which will depict a crystal. I am not allowed to stack them any way I like, because I must continue to satisfy the cosmological principle.

One way to achieve this is to stack the planes of lattice points so that the spacing between adjacent planes is always the same, equal to *c*, and further so that each point in a plane is directly above a point in the plane below (you might see some similarity to the way we stacked the one-dimensional lattices to get a two-dimensional rectangular lattice).

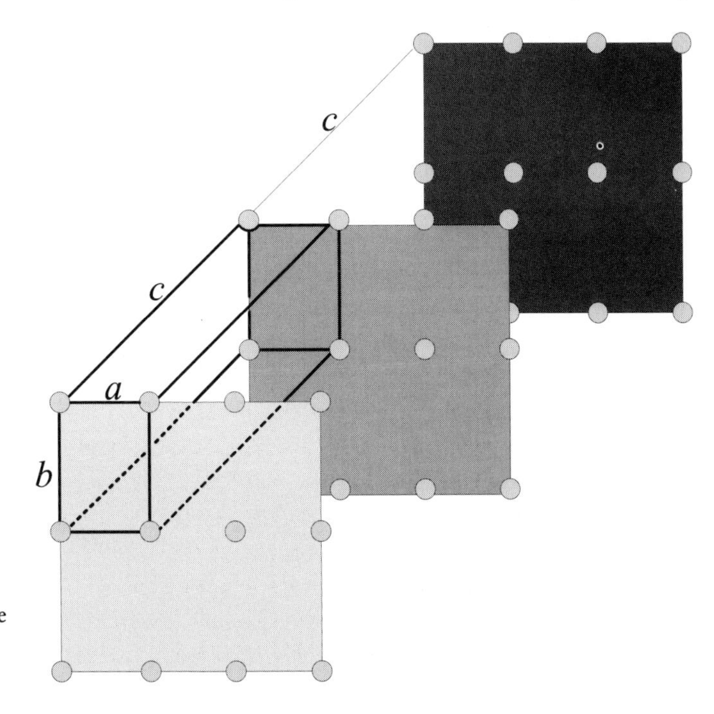

FIGURE II-12. The generation of a three-dimensional lattice using two-dimensional rectangular lattices. The planes are stacked at equal distances apart so that the lattice points in different planes lie directly over one another. The result is called a simple orthorhombic lattice. The unit cell is outlined with heavy lines, shaped like a brick with edges *a*, *b*, and *c*, and is shown again in fig. II-13, part (1).

I end up with a three-dimensional lattice. Figure II-12 shows you what a small portion of this lattice looks like. I think you can see what I meant about the difficulty of drawing a picture of a three-dimensional object on a two-dimensional page. However, if you use your imagination a bit, you can see that the lattice is made up of a lot of tiny bricks each with edges a, b, and c. I have marked one of them at the top left with heavy lines to help you visualize it. This tiny 'brick' is the *unit cell* of the lattice, and is shown in fig. II-13, labelled (1). I can build up the whole lattice out of these unit cells. The positions of the atoms surrounding any given atom in the lattice satisfy the cosmological principle.

There is another way of stacking rectangular lattices so as to satisfy the fundamental condition. I place the second lattice over the first at a distance $c/2$, with each atom lying exactly over the centre of a little rectangle in the first lattice. I place the third lattice at a

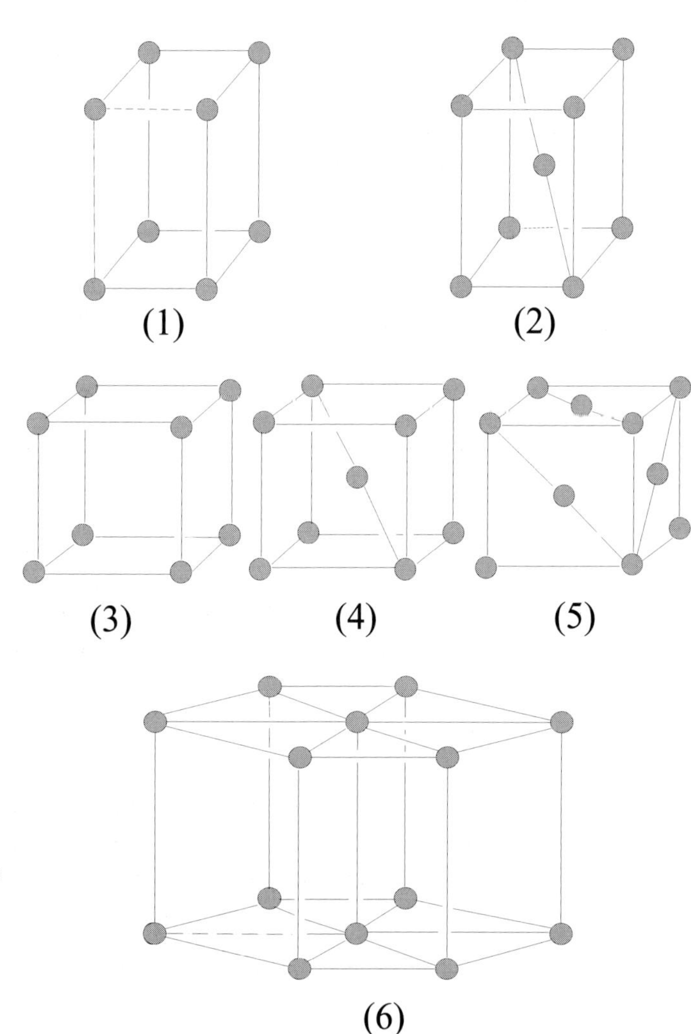

FIGURE II-13. Unit cells for some three-dimensional lattices: (1) simple orthorhombic; (2) body-centred orthorhombic; (3) simple cubic; (4) body-centred cubic; (5) face-centred cubic (only the three front faces are shown for clarity); (6) simple hexagonal. The unit cells (3), (4) and (5) are shaped like cubes. The lattice represented by (6) results from a vertical stacking of the two-dimensional triangular lattice.

distance *c* from the first, but with each atom now lying exactly over an atom of the first lattice (there is a similarity to the way we made a centred rectangular lattice in two dimensions). I show the resulting unit cell in fig. II-13, labelled (2). This lattice is different from the previous one, but still satisfies the cosmological principle. More three-dimensional lattices can be assembled using each of the other four two-dimensional lattices.

QUESTION Is one allowed to take a mixture of several kinds of two-dimensional lattices to build a three-dimensional lattice?

ANSWER No, because then the cosmological principle will obviously not be satisfied.

Even with this constraint, you might think that the possibilities are practically limitless. There are however only fourteen lattices with different symmetries that I can build satisfying the cosmological principle. Figure II-13 shows the unit cells and the names of some of these lattices. The circles in these figures represent the lattice points, and the lines joining them are there to guide the eye in seeing the shapes of the unit cells.

We thus come to the astonishing conclusion that the well-nigh infinite varieties of crystals that occur in nature or are made in the laboratory belong to one of just fourteen crystal lattices. I recapitulate the line of thought that has brought us to this conclusion:

1. All atoms of a given element are identical.
2. As a consequence, the cosmological principle holds for a solid.
3. A lattice is characterized by its symmetries.
4. Two lattices are different if they do not share exactly the same set of symmetries.

Nature confirms our conclusion: all crystals that occur naturally or are made in the laboratory belong to one of these fourteen lattice types.

We talked earlier of a number of symmetries and other characteristics of two-dimensional lattices. Suitably modified, they exist in the three-dimensional lattices too, but it is not always easy to see them in a picture of the lattice on a page. Physicists, who have the same difficulty in visualizing these lattices, resort to models assembled from balls and rods. All these properties can be proved by applying mathematics. I list the results below:

1. There is translational symmetry for displacement from one lattice point to any other lattice point.
2. A straight line connecting any two lattice points, when extended, will pass through other equally spaced lattice points, and only those points.
3. A lattice can be thought of as being made up of identical parallel planes in more ways than one, each with its own characteristic spacing within and between planes.

By analogy with the discussion of how a two-dimensional crystal grows so that it is bounded by straight lines of close-packed atoms, one can understand why natural crystals are bounded by plane faces, each a plane of close-packed atoms.

FIGURE II-14. Part (*a*) shows the graphite structure, which is a triangular lattice with pairs of atoms like A and B attached to each lattice site. Part (*b*) shows the cubic unit cell for a crystal of common salt (sodium chloride).

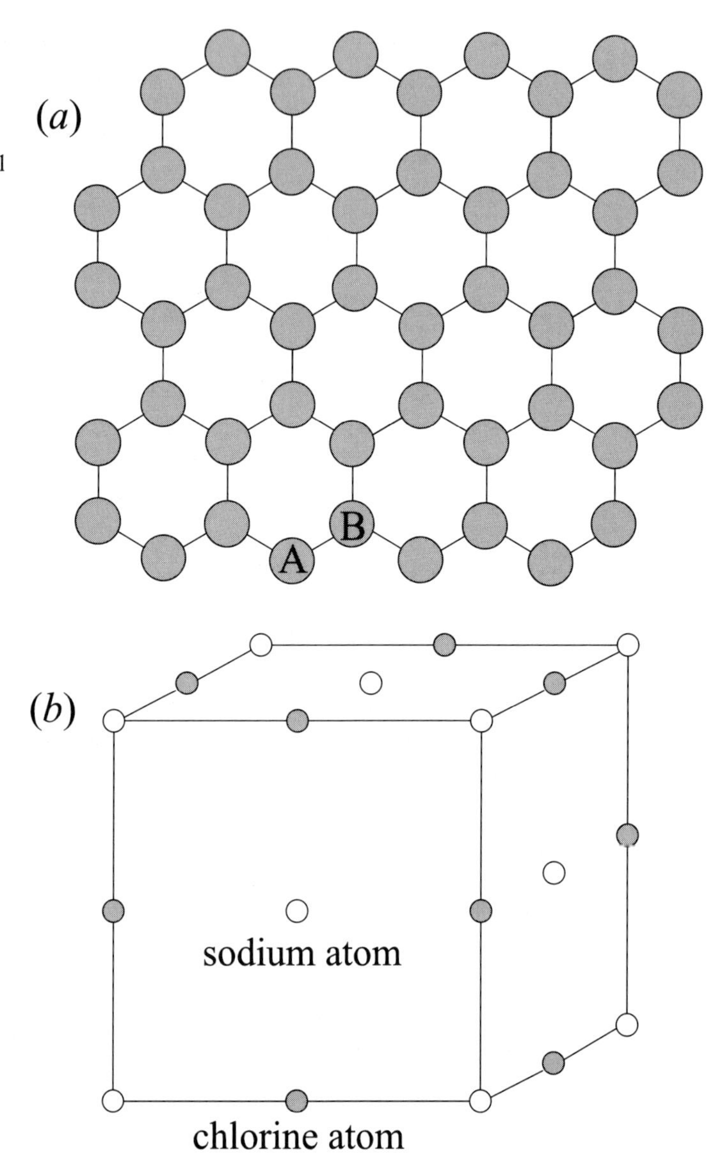

8 Crystal structures

The lattices we have been considering are not real objects, because there are no atoms in them. They are just sets of points, distributed in space in a way that satisfies the cosmological principle. It is only after we associate atoms with the lattice points that we get a physical object. A crystal of silver has the atoms sitting at the points of a face-centred cubic lattice, whose unit cell is shown in fig. II-13, labelled (5). We say that silver crystallizes in the *face-centred cubic structure.* Brass is an alloy in which 65 per cent of the atoms are copper and 35 per cent are zinc. A crystal of brass also has a face-centred cubic structure, in which the copper and zinc atoms occupy the lattice sites at random.

QUESTION In such an alloy there are two kinds of atoms, and so the cosmological principle cannot be satisfied. What happens to the lattice spacing?

ANSWER Call the two kinds of atoms A and B. Because an A atom is different from a B atom, the lengths and also the strengths of the A-A, A-B and B-B springs will be different from one another. These differences are usually small. So the crystal of the alloy will have its atoms displaced, but only slightly, from the exact lattice sites.

Yet another kind of crystal is made up, not of single atoms sitting at each lattice site, but rather of a group of two or more atoms associated with each site. Further, the atoms in the group could all be the same, or they could be different from one another. We could also have molecules around each lattice site. I give below examples of these different possibilities.

The element carbon exists in nature with two different crystalline forms, as graphite and as diamond. Graphite consists of identical planes, one of which is shown in part (*a*) of fig. II-14. The picture does not look at all like one of the five two-dimensional lattices. It is in fact a triangular lattice, with pairs of atoms like the ones marked A and B attached to each lattice point. This arrangement of atoms is known as the graphite structure.

My next example is common salt, which consists of equal parts of the elements sodium and chlorine. The sodium atoms form a face-centred cubic lattice. There is one chlorine atom for each sodium atom, sitting on its right at a distance $a/2$, where a is the length of the cube edge. I show in part (*b*) of fig. II-14 the resulting unit cell for a crystal of common salt. This is called the rocksalt structure, after the mineral form of common salt. You may have seen large cubic crystals of rocksalt in museums of science. Figure II-15(*a*) shows what common salt looks like through a microscope. You can see the cubic nature of the grains, arising from the cubic unit cell.

My last example is a crystal composed of sugar molecules. I show in fig. II-15(*b*) a photograph of grains of sugar made with a microscope. Each grain is a crystal. The molecule has six atoms of carbon, six of oxygen, and twelve of hydrogen. It is somewhat more complicated than the examples we have looked at so far. Nevertheless, the crystal still forms one of the fourteen lattices, the orthorhombic lattice, whose unit cell is like a brick with unequal edges. The flat faces of the crystals are sheets of molecules lying relatively close to one another.

The crystals in fig. II-15 do not look perfect: they have rounded edges, and some are broken. This is because of the handling they have received and their solubility in water. Note particularly that while their sizes vary, the basic shape is the same for all crystals of each of the two substances. This is a beautiful manifestation of the underlying orderly arrangement of the atoms or molecules in the crystal.

I have so far used symbols like a, b, and so on to denote the quantities that describe a

FIGURE II-15. Photographs of two crystalline substances: (*a*) common salt; note the cubic shape of the crystals, and (*b*) sugar; the crystal structure is orthorhombic, and each of the plane surfaces of the individual crystals is a close-packed plane of sugar molecules. (Courtesy of Frau Gabrielle Goerblich, Walther Meissner Institute, Garching, Germany.)

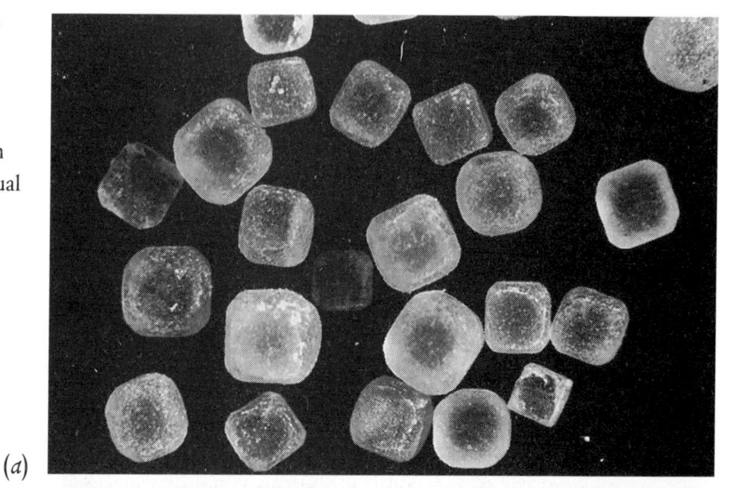

(*a*)

(*b*)

crystal. It is useful to know the order of magnitude of these quantities in real crystals. They of course vary from crystal to crystal, but not by a great deal. Suppose I take a crystal in the shape of a cube whose edge is one centimetre long. It will contain about 10^{21} atoms, which means about 10^{7} atoms along an edge and 10^{14} atoms on a face. A little more arithmetic will show you that the spacing between atoms is about 10^{-7} centimetre. There is nothing in our everyday experience that helps us visualize such tiny lengths and huge numbers, but they are nonetheless real in the crystal. Fortunately, such visualization is not necessary to proceed with our story.

9 Glassy solids

We saw earlier that if we cool a melt fast enough, the resulting solid could be glassy, and the atoms (or molecules) will not sit in an orderly crystalline arrangement. This seems at first glance to violate the reasoning we have used to explain the formation of crystals. A little further thought shows that there is in fact no problem. Remember the virtual springs that we imagined connecting each atom to its surrounding atoms? The very rapid cooling leaves the atoms at or very near the positions they occupied in the liquid, each feeling no net force so that we still have a solid. However, the springs surrounding each atom do not all have the same relaxed length that they would have had in the crystal. Some will find themselves a little compressed, others a little extended, but altogether in such a way that there is still no net force on each atom. When we realize that each atom has many springs connecting it to the surrounding atoms, we can see that there is a good chance that the atoms can find positions, not ordered, in which each atom feels no force even though each spring is a bit compressed or extended. The result is a glassy solid. This picture also helps us understand why some solids are more easily formed in the glassy state than others: the exact details of the 'springs' play a crucial role. We recall that a glass, taking a long time that may be several centuries or more, will change to a crystalline form, which therefore appears to be the preferred form for the atoms in a solid to take.

10 Summary

We saw in section 1 that as a liquid is cooled to form a solid, the result is a single crystal, or a polycrystal, or a glassy solid, depending on the cooling rate. The atoms sit in an orderly arrangement in the crystal and polycrystal, but not so in the glassy solid. The arrangement in which the atoms prefer to be is the crystalline one.

The cosmological principle as it applies to crystals is introduced in section 2. It is seen as a consequence just of the fact that all atoms of a given element are identical to one another. It seems reasonable therefore that the principle should hold for the preferred atomic arrangement in the solid.

The simplest solid we can think of is just a line of atoms. If this line is to satisfy the cosmological principle, the atoms will be equally spaced along a straight line. This is the one-dimensional crystal, which is discussed in section 3.

In section 4, the five possible two-dimensional lattices are built up from a collection of identical one-dimensional lattices.

Section 5 makes precise the idea of order in a crystal by introducing the concept of symmetry as it applies to crystals. The symmetries of two-dimensional lattices are discussed in section 6. Several kinds of symmetry are seen to exist: of translation, of rotation, and of reflection.

Stacking a number of two-dimensional lattices on top of one another while maintaining the cosmological principle leads to fourteen different three-dimensional lattices. Section 7 introduces them and some of their characteristics.

In section 8, several ways are described in which atoms can be associated with lattice sites, leading to crystals of materials each with its characteristic structure.

Section 9 deals very briefly with a glassy solid, and points out that after a long time of possibly centuries, it will have changed to a crystalline form, which seems to be the preferred form of atoms in a solid.

This concludes our description of crystals. The crystals we have assembled have no properties other than their symmetries. Their physical properties arise only when we put real atoms at the lattice sites. Before doing that, we need to learn about an isolated atom in terms of its nucleus and electrons in some detail. We shall then be able to see what happens to the atom when it finds itself in a crystal. Finally we shall use the resulting picture of the crystal to explain various properties of a solid.

I have said that an atom consists of a nucleus with some electrons around it. To describe it in more detail, I need to introduce the description of nature called *quantum mechanics.* This is the subject of chapter III.

III PARTICLES AND WAVES

1 Some preliminaries

We have, until now, thought of atoms as tiny hard spheres that feel a force of attraction towards one another. We ignored the fact that an atom consists of a nucleus surrounded by one or more electrons. If we stopped at this, we would make little progress towards understanding the properties of solids, other than that the atoms may form crystals. We must therefore go on and take a look at the internal structure of the atom itself, and see how this changes when the atom is surrounded by other atoms in a crystal. We shall then find that we need a description that is somewhat removed from our everyday experience. There is a name for this description, *quantum mechanics*: mechanics, because it deals with the motion of objects, and quantum, because features of the motion, such as energy, are *quantized*, meaning that they can only have any one of a set of distinct values, and no other value in between. Imagine for example a car which is so constructed that it can travel only at speeds of 10, or 50, or 100 km per hour, and not at any other speed. Then we could say that the speed of the car is quantized. This idea of quantization is something alien to our usual way of thinking. After all, a car is able to go at any speed we like up to its maximum speed. We shall find that there is no such freedom of choice for electrons and nuclei. In this chapter I shall introduce you to the essentials of quantum mechanics, and then in the next chapter show you the description of an atom that it provides. I have summarized the content of this chapter in section 13 at the end (page 61). You may want to read it now, so that you have a preview of what lies ahead.

2 Motion

As I write this, I take a moment to look out of the window. I notice that there is a breeze that is causing the leaves on the trees to flutter. Birds are flying in the air. I see a stream flowing among the trees, and there are small waves on its surface that are themselves also in motion. Although I cannot see them, I know that the molecules in the air around me are in a perpetual state of motion. The paper clip on the table before me seems to be sitting there, doing nothing. If I look at what the atoms in it are doing, using the right sort of microscope, I find that they are all jiggling a little bit about their regular crystalline positions. It would appear that the world around us is always in various kinds of motion. At first glance, there seems to be a daunting variety of such motions, and of moving things. However, the essential features of each of these motions can be assigned to one or the other of just two types, namely the motion of *particles*, or of *waves.*

Before proceeding, I should make a general observation. I use words in this book, many of which are taken from everyday usage, to denote certain concepts. When so used, they have very narrowly defined meanings, and should not be confused with their everyday connotations. When I say *crystal,* I mean a solid with an orderly arrangement of its atoms as described in chapter II, and not an expensive piece of glassware. Other examples are *energy, momentum, force, particle, conservation.* As Humpty Dumpty said in Lewis Carroll's *Through the Looking-Glass,* 'When I use a word, it means just what I choose it to mean, nothing more nor less.' The glossary beginning on page 238 lists these words with explanations.

We already have some idea of what is meant by a wave; we have all seen waves on the surface of a stretch of water, for example. A wave is spread out in space and moves in some direction. It has many alternating crests and troughs.

I now explain what I mean by a particle. Here is a definition: *A particle is an object that has a mass and occupies a vanishingly small volume.* As I make the volume smaller and smaller, it finally becomes a point in space. I can therefore think of a moving particle as a mass that is concentrated at a point and is moving about. A particle is thus an idealization of an ordinary moving object. It is in this sense that I say that electrons and nuclei are particles. An essential feature of a particle is that it is at some point in space at each instant of time: *it is localized in space and time,* in contrast to a wave, which is *spread out in space at every instant of time.* Figure III-1 illustrates the difference between a particle and a wave.

Electrons and nuclei have three other properties in addition to their mass: they have an electric charge, they behave as if they were spinning about an axis, and they respond to a bar magnet the way a magnetic compass needle does, as if they were tiny bar magnets with a north pole and a south pole. Furthermore, these properties remain exactly the same at all times and places. There is nothing we can do which will change the mass, the charge, the spinning, or the magnetism of an electron or a nucleus. They are what lead ultimately to our understanding of the properties of matter.

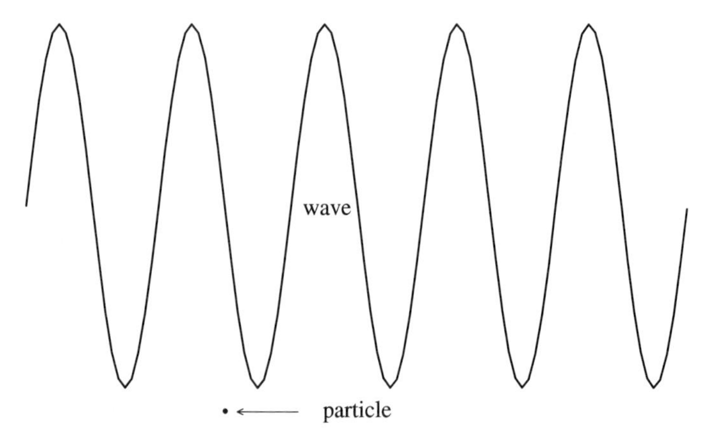

FIGURE III-1. To illustrate the difference between a particle, which is very localized in space, and a wave, which is extended over all of space.

3 Particles

To begin with, a particle has mass, say M grams. If it is an electron, then the mass is 9.11×10^{-28} gram. For a proton, it is 1.67×10^{-24} gram, or about 1800 times the mass of the electron. The mass of a particle is the measure of the amount of matter it contains. It is the property which gives it its heaviness, its *weight*, due to gravity, and also determines how it moves when a force is applied to it.

As the particle moves about, it is at some point in space at each instant of time: it is *localized in space and time.* The particle moves in some *direction* with some *speed.* In our everyday experience with moving objects, we are often content with knowing the speed, so many kilometres per hour, for example, and do not explicitly state the direction of motion. We know that direction can often be important too: we would not easily get to Cambridge from Oxford or to Cleveland from New York if we insisted on travelling south all the time. Similarly, when we try to understand the nature of matter in terms of its basic constituents (atoms, electrons, etc.), we use a quantity called *velocity* denoted by the symbol $\mathbf{v}$ (in heavy type) which tells us not only how fast but also in which direction. When I say that a particle has a velocity $\mathbf{v}$, I mean that it has a speed of v in specified units (centimetres per second, for example) in some specified direction. Two particles moving with the same speed but in different directions have different velocities, as also two particles moving in the same direction but with different speeds. We call quantities that have both a magnitude (so many centimetres per second in the case of $\mathbf{v}$) and a direction, *vectors*, and denote them in heavy type. When I want to speak only about the magnitude of the vector, I shall use the same letter but now printed in ordinary italic type, v in the case of velocity. A quantity that has only a magnitude and no direction is called a *scalar*. Thus an interval of time, t seconds, is a scalar.

In solids we are concerned not only with the motions of the atoms, but also with how they influence one another. A simple example will illustrate the point. In air, say, each molecule is whizzing about, colliding every now and then with another molecule. After each such collision, the two molecules have velocities that are different from what they were before. The final velocities will depend not only on the initial velocities of the two molecules, but also on their masses. I would certainly notice the difference between a fly and a cyclist colliding with me as I am walking along a path, even if they both were travelling at the same velocity. So, in order to consider collisions (or more generally *interactions*), between particles, I need a concept, and a symbol for it, which combines mass and velocity. I call it *linear momentum*, and give it the symbol $\mathbf{p}$ (again a vector). Its direction is the direction of the velocity $\mathbf{v}$, and its magnitude p is given by the mass M multiplied by the speed v. In short, we can write

$$\mathbf{p} = M\mathbf{v}, \text{ a vector equation,}$$

and

$$p = Mv, \text{ a scalar equation.}$$

Note that the second equation gives only the magnitude of the vector $\mathbf{p}$.

There is another kind of motion possible for a body, in which it is spinning about an axis, but otherwise not going anywhere. The body as a whole has zero velocity, and therefore zero momentum. All its atoms however are spinning around in circles about the axis. The momentum connected with this motion is called *angular momentum.* We note that it is essentially different from the linear momentum we got by multiplying the mass and the velocity. Angular momentum is also a vector, and its direction is along the axis of rotation. Its magnitude is fixed by how fast the body is rotating and how its mass is distributed around its axis of rotation. We recall that electrons and nuclei behave as if they were spinning about an axis: they have an angular momentum that is an intrinsic property. In what follows, I shall simply use 'momentum' when the context makes it clear which kind is meant.

We have one final concept to introduce, that of *energy.* Here is another of those words of normal speech, which we use in this book with a very specific meaning. I start by defining energy, and then give some examples.

A body has energy either if it is already in a state of motion, or if, even though at rest, it can through appropriate means put itself or another body into a state of motion.

In the first case, we say that the body has *kinetic energy*, and in the second case, *potential energy.* The kinetic energy is denoted by the symbol K, and is given by

$$K = \frac{Mv^2}{2}$$

or equivalently

$$K = \frac{p^2}{2M}$$

as can be seen from our definition of p as equal to Mv. If a mass of 1 gram is going at a speed of 1 centimetre per second, it is said to have a kinetic energy of one-half *erg.* If the mass is travelling at 2 centimetres per second, its kinetic energy is 2 ergs. Note that doubling the speed quadruples the energy: cyclists and car drivers should be aware of this. Energy has just a magnitude of so many ergs, and no direction attached to it, unlike momentum: energy is a scalar.

QUESTION Can you name another scalar quantity?

ANSWER The mass M. It has only a magnitude of so many grams, and no direction.

I now turn to the second type of energy, *potential energy.* It is a form of energy that, under the right conditions, can be converted into kinetic energy. I illustrate it with a few examples. I place a marble at the edge of a table. It is not moving, and therefore has no kinetic energy. I now give it a gentle shove to send it over the edge. It starts falling, with ever increasing speed, until it hits the floor: it has acquired kinetic energy. I say that the marble sitting on the table has potential energy, because when I remove the table's support the marble begins to move. Since this property of the marble arises from gravitation, I call it *gravitational potential energy.* Gravity acting on the marble causes it to move when it is no longer held up by the table. You will note that even when the ball was sitting still on the table, it had the *potential* to move; hence the name potential energy. As another example, consider an electric battery. I connect it to a light bulb with a pair of wires and a switch. When I close the switch, the battery drives a stream of electrons – we call it an electric current – through the bulb causing it to light up. The electrons move because of the battery. I say that the battery has *chemical potential energy*. If I replace the electric battery with a solar battery and shine sunlight on it, I get the same result. Thus sunlight also has energy that can be converted to the motion of particles, electrons in this instance. This energy is commonly called solar energy. A physicist would probably call it *electromagnetic potential energy*, because light is intimately connected with electricity and magnetism, as we shall see later. Yet another example: if I stretch a rubber band between my hands and let go, it flies off. The stretched band has *elastic potential energy* when it is stretched, and this is converted to kinetic energy when I let go the band. If it is clear from the context that I am talking about some particular form of potential energy, then I may simply say gravitational energy, chemical energy, and so on.

QUESTION Can you think of an example that shows that heat is also a form of energy?

ANSWER A steam locomotive. Energy in the form of heat is converted to the kinetic energy of motion of the locomotive.

Finally, I give an example that is probably known to everyone: Albert Einstein's discovery that a particle of mass M has a potential energy by virtue of its mass alone, equal to Mc^2 where c is the speed of light, namely 3×10^{10} centimetres per second.

The foregoing does not exhaust all possible forms of energy, but is meant to show that it comes in many different forms. The examples also illustrate the fact that it is possible to convert the energy from one form to another: the locomotive converts thermal energy to the energy of motion. Now it often happens that one and the same particle can have energy in more than one form at the same time. For example, a particle sliding on a tabletop has both kinetic energy and gravitational potential energy. In such cases, we add up all the separate energies and call it the total energy E. We have

$$E = \text{kinetic energy} + \text{gravitational potential energy} + \text{etc.}$$

I summarize now the essential characteristics of a particle: a particle is localized in space and in time, and it has a mass M, a momentum $\mathbf{p}$, and a total energy E.

We have so far considered a single particle. The pieces of matter whose properties we wish to understand consist of many particles such as atoms and molecules, themselves composed of nuclei and electrons. If I take a piece of matter, I get its total energy by adding up the individual energies of each of its constituent particles, and its total momentum by doing the same with the individual momenta (plural of momentum). I have no problem doing this with the energies, because they are scalars, just numbers. How do I do this with momenta, which are vectors with both magnitude and

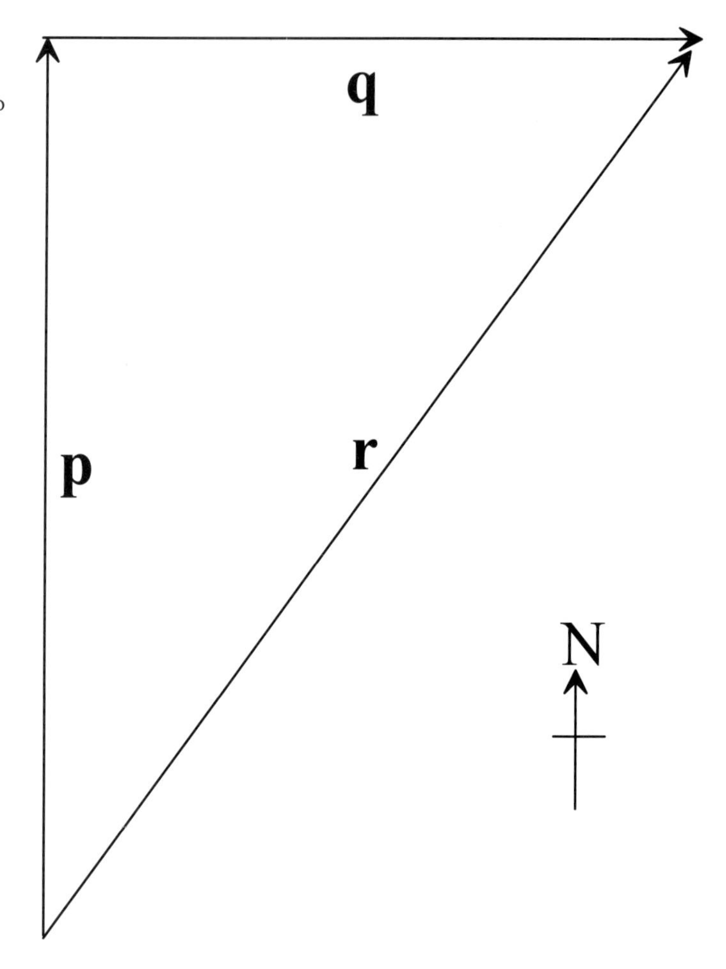

FIGURE III-2. Adding two vectors together: **p** added to **q** results in the vector **r**. This procedure is applied to the question about the canoe on the river. The direction of north (N) is shown.

direction? We do this as shown in fig. III-2 for vectors **p** and **q**, which for example could be two velocities. I draw an arrow whose length and direction are the magnitude and direction respectively of **p**. I draw a similar arrow for **q**, with its tail at the head of **p**. Then the sum of the two, which I call **r**, in magnitude and direction is given by the arrow which connects the tail of **p** with the head of **q**. The resultant velocity is then **r**.

QUESTION I paddle north in a canoe at 4 km/hour (**p**) across a river that is flowing east at a speed of 3 km/hour (**q**). What is my actual velocity as seen by a person standing on the bank of the river?

ANSWER The canoe is going north in the water at 4 km/hour, and the water itself is flowing east at 3 km/hour past the person on the river bank. Using the above prescription and direct measurement, or the theorem of Pythagoras and trigonometry, I find my velocity to be 5 km/hour (**r**) in a nearly northeasterly direction, 37° from north to be precise, as seen by the person on the river bank. Figure III-2 illustrates this.

4 The conservation of energy and momentum

The subject of this section is a profound and beautiful basis of all of physics. These conservation principles are sometimes explicitly, and often implicitly, contained in much of what follows in the book. I shall first state what is meant by the conservation of energy and momentum:

The total energy, the total linear momentum, and the total angular momentum of a closed system are constant at all times: they are said to be conserved.

The statement summarizes three fundamental principles that govern all of physics: the conservation of energy, the conservation of linear momentum, and the conservation of angular momentum. When we say 'conservation of energy' here, we mean something different from switching lights off when they are not needed. We should of course be as economical as possible in the use of energy in our daily life. When we talk here about energy being conserved, it is not a matter of personal choice but a fact of nature. By 'closed system', we mean some portion of our surroundings that is so fixed up that neither matter nor energy can enter it or leave it. This is an idealization of what can be achieved in the real world, where it not easy to exclude totally a flow of energy and/or matter. Some hot coffee in a closed thermos flask is an approximation to a closed system. No form of matter can enter or leave the flask, nor much energy as heat or light. It is not a perfect closed system: the coffee does cool eventually, because it slowly loses thermal energy to the outside.

Let us see how the conservation principles apply to the case of a ball falling off the edge of a table and coming to rest on the floor. At the start of the fall, the ball has only gravitational potential energy. During the fall, the potential energy is continually converted to kinetic energy of motion, in such a way that the total energy remains constant. Just before the instant when the ball hits the ground, its initial potential energy has been completely converted to an equal amount of kinetic energy, so that the total energy is still the same. After impact, the ball is at rest on the floor, and both its kinetic and potential energies are zero. Figure III-3 shows the three successive stages. You may wonder what happened to energy conservation when the ball came to rest on the floor, since both its kinetic and potential energies then are zero. Not to worry: the impact makes a sound and the ball warms up a bit, because the total energy has now appeared partly as sound and the rest as heat. If you have ever hammered a nail into the wall and noticed the nail warming up, you will have seen the conversion of kinetic energy into thermal energy. But how about the linear momentum? The ball started from rest with zero momentum, and continually increased its speed as it fell. Its momentum certainly did not remain constant. Then we note that the closed system is not just the ball by itself, but the ball plus the earth towards which it is falling, as seen by an extraterrestrial creature somewhere out in space. This creature sees the earth move up to meet the ball with a momentum which at each instant of time is exactly equal to and opposite to the momentum of the ball, so that the total momentum at all times is zero: it is conserved.

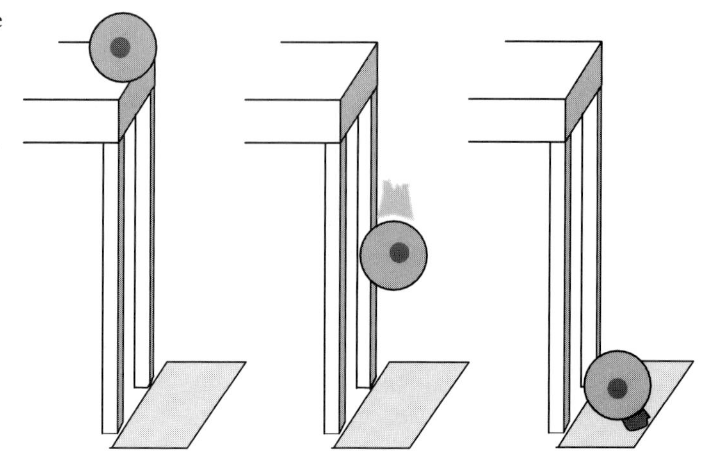

FIGURE III-3. From left to right, three successive stages of a ball falling off a table. The initial gravitational potential energy is converted to kinetic energy and then to the energy of sound and heat.

QUESTION Then should I not include the earth in considering the energy also?

ANSWER Yes, you should. However, because of the enormous difference in masses of the two objects, the kinetic energy of the earth is absolutely negligible compared to that of the ball, even though their momenta have equal magnitudes. You remember that the kinetic energy is equal to the square of the momentum divided by twice the mass. So the kinetic energy of the earth is small compared to that of the ball in the same proportion as the mass of the ball is small compared to that of the earth.

As mentioned in chapter II, these conservation principles arise out of the fundamental symmetries of space and time. To go deeper into this will lead us too far from our main course, to which we must now return.

5 Waves

We encounter waves in nature in a great variety of forms: the ripples which spread out from a stone dropped into a pool of water, sound waves in air, light which is an electromagnetic wave that travels through even empty outer space, and great ocean waves, to name a few. We have a similarly great variety of moving objects in nature, and in order to distil out the essential aspects of their motion, we came up with the idea of a *particle.* We shall now similarly consider an idealized wave, shown in fig. III-4 at five successive instants of time. To be specific, imagine it to be a wave on the surface of water.

The wave, unlike a particle, is not localized, but is spread over a region of space. It consists of alternating crests and troughs, moving with a certain speed. By including the direction in which the wave is travelling, I can talk about the *velocity* of the wave. Although the wave itself is travelling in a certain direction, there is no general motion of water in that direction. I see this by floating a piece of cork on the surface of the water. As the successive crests and troughs go by, the cork moves up and down, but is not carried forward. The wave does not carry matter along with it, whereas the motion of a particle is certainly the motion of matter. The maximum distance that the cork moves from its mean position, as it oscillates up and down, is called the *amplitude* of the wave and denoted by the symbol A. At any fixed point along its path, the wave is oscillating up and down f times per second, and f is called the *frequency* of the wave. The distance between adjacent crests is always the same, and equal to the distance between adjacent troughs. This distance is called the *wavelength*, and denoted by the Greek letter λ (lambda).

Figure III-4 illustrates some of these points. You can see that this wave has a wavelength of 1 metre. The broken vertical line near the left end of the picture intersects the wave at the position of the cork at successive instants of time. It is on a crest at 0 second, has moved down to a trough at 2 seconds, and is back again on the following crest at 4 seconds. It has completed one oscillation in four seconds, and therefore the frequency is $\frac{1}{4}$ oscillation per second. Each crest is moving to the right as time elapses, and has moved a distance of 1 metre in 4 seconds. The speed of the wave is therefore $\frac{1}{4}$ metre per second.

QUESTION Can you see that, if the speed of the wave is v, then v is equal to the product of f and λ?

ANSWER I look at the cork, and find that f crests are passing it each second. The crests are spaced a distance λ apart. Therefore the crest, which one second ago was at a distance f times λ, is now passing the cork. So the speed of the wave is f times λ centimetres per second. I can write this as $v = f\lambda$.

If I know the wavelength and frequency of a wave, I just multiply the two together

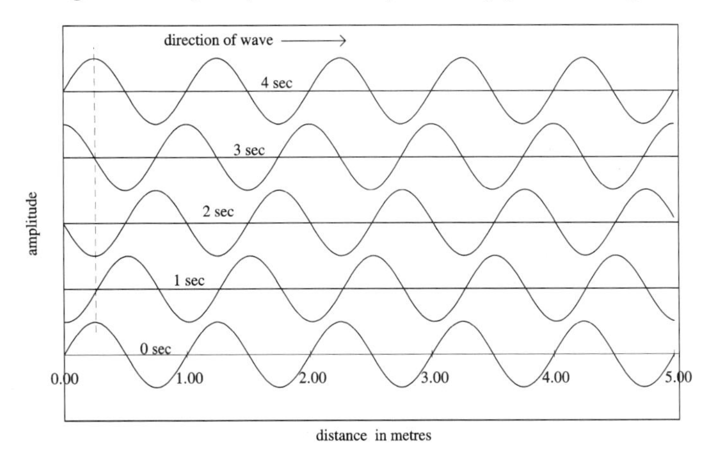

FIGURE III-4. A wave travelling to the right, as seen at successive intervals of one second. The intersection with each wave of the broken vertical line near the left indicates the position of the cork referred to in the text. You can see from the figure that there is one oscillation in four seconds (or a frequency of one-fourth oscillation per second), a wavelength of one metre, and a speed of one-fourth metre per second.

to get the speed. This speed depends on the nature of the wave and the substance in which it travels. The speed of sound waves in air is 331 metres per second, in water 1480 metres per second. The speed of light waves in free space is 3×10^8 metres per second, in water 2.24×10^8 metres per second.

The last three attributes of wave motion that we have described, namely the amplitude, frequency and wavelength, have no counterparts in particle motion. In addition, I see no obvious way of talking about the momentum or energy of the wave, as I could with a particle. I know however that waves also do have energy and momentum, as is shown by the following effects with sunlight, which is a form of electromagnetic waves. The energy of sunlight is converted into the energy of motion of electrons in the current produced by a solar battery. The momentum of sunlight causes the tails of comets to stream away from the sun. We shall see later how quantum mechanics tells us how to associate energy and momentum with waves.

Since water consists of molecules (each made up of two atoms of hydrogen and one of oxygen), I could argue that in principle I could describe water waves in terms of the motion of water molecules: a task which would need the lifetime of the universe to complete – perhaps. Then there are electromagnetic waves in the form of light, which travel through empty space where there are no molecules or other forms of matter at all. Light must therefore be a pure wave with nothing in the form of particles of any sort associated with it. So it would seem that we must accept particle motion and wave motion as two distinct and non-overlapping forms in nature.

I now come to yet another difference between particle motion and wave motion. When two particles are moving together, I have no difficulty in thinking of the motion of each particle independently of the other. If the two collide with each other, they would continue their separate motions afterwards, still clearly distinguishable from each other. If on the other hand I get two waves to come together, something very dramatic happens, as illustrated in fig. III-5. I take two waves, which I label (1) and (2), with slightly different wavelengths. I let them combine so that at each point the new amplitude is the sum of the amplitudes of the two waves, and get a resulting wave that I have labelled 'sum'. This wave has an amplitude that waxes and wanes

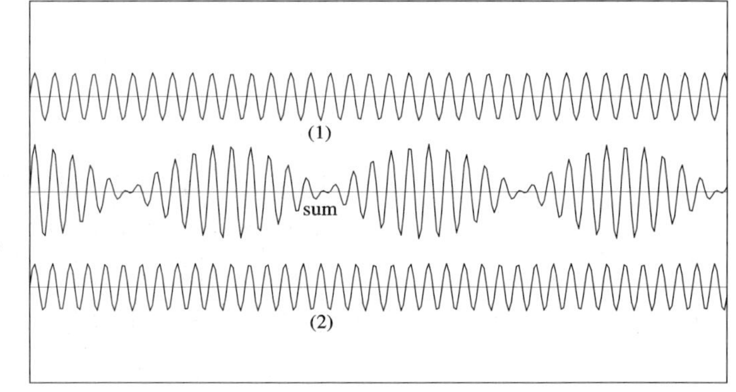

FIGURE III-5. The interference of two waves travelling together. The separate waves labelled (1) and (2) add up to give the wave labelled 'sum'.

regularly, unlike its two constituent waves. Because their wavelengths are different, the two waves find themselves successively in step and out of step. When they are in step, they add up to give a larger amplitude. When they are out of step, they cancel each other to result in zero amplitude. I say that the two waves interfere with each other, and call the phenomenon *interference.* The places where the amplitude is practically zero are called nodes, and where it is a maximum, antinodes. If I looked only at the nodes, I would think there was no wave at all. The wave seems to exist only between the nodes.

I made fig. III-5 with a computer. Figures III-6 and III-7 are photographs showing the interference of real waves on the surface of a lake. Figure III-6 is a view of the circular waves produced by dropping two stones into the water at the same time about two metres apart. Figure III-7 shows the region where the two waves overlap. The interference pattern produced by the two waves is clearly visible.

FIGURE III-6. Two circular waves starting at different points on the surface of water in a lake. Note the interference between the waves where they overlap.

FIGURE III-7. A close-up view of the region of interference of two waves similar to the ones in fig. III-6. Crests and troughs from the two waves reinforce each other, whereas the crest of one is cancelled by the trough of the other.

There is still another difference between particles and waves. If I place an obstacle in the path of a moving particle, the particle may either hit the obstacle and change its subsequent motion, or miss it and continue on its way as if the obstacle did not exist. On the other hand, the motion of the wave after it passes an obstacle is quite different from what it would have been without the obstacle. Put differently, a particle cannot go through an obstacle, but a wave goes around it and continues although with an altered form. Waves undergo *diffraction* by an obstacle, whereas particles do not. When I go to an open air concert, I sometimes find a very large person sitting in front of me, pretty much blocking my view of the stage. Even though there are no walls or roof to reflect the music towards me, I can still hear it almost as if the person were not there. Since light is also a form of wave motion, you may wonder why the light from the stage is also not diffracted round the person to reach me, so that I can still see the stage. The answer lies in the difference in wavelength between the two kinds of waves: in the range of centimetres to metres for sound, and a few times 10^{-5} centimetres for light. The size of the person in front of me is comparable to the wavelength of sound, but is enormous compared to the wavelength of light. When the size of the obstacle is comparable to or smaller than the wavelength, we get strong diffraction effects. If the size of the obstacle is very much greater than the wavelength, the wave is blocked by the obstacle.

QUESTION If you hold a compact disc so that light from a window is reflected from its playing surface into your eyes, you will see a beautiful spectrum of colours. What causes this?

ANSWER There is a regular pattern of dots on the surface, with a spacing that is comparable to the wavelength of light. So the daylight is diffracted, and different wavelengths (which correspond to different colours) are sent in different directions. The result is the spectrum of colours which you see. It is this pattern of dots, invisible to the naked eye, which is 'read' by the compact disc player and finally emerges as music.

6 Particles and waves: a comparison

I summarize in table III-1 the characteristics of particles and waves, so that the differences between them can be seen clearly. This table shows us that in the everyday world, a particle and a wave are very distinct entities. Each has certain characteristics that are not possessed by the other. If I point out to you something that is moving, you are able to tell me without hesitation whether it is a particle or a wave. A ball thrown up in the air is a particle, ripples on water are a wave. You will think that I am somewhat odd if I tell you that the ball can sometimes behave like a wave, and the ripples like particles. Well, that in essence is what I am about to tell you!

Table III-1. *Properties of particles and waves*

particles	waves
localized in space and time	not localized in space at any time
mass, M	cannot imagine mass of a wave
momentum, p	wavelength, λ
energy, E	frequency, f
no interference	interference
no diffraction	diffraction

7 The electron: particle and wave

We want to understand the properties of matter in terms of the constituent atoms, which are themselves made up of electrons and nuclei. To be specific, I talk about electrons here. I describe three experiments that can tell me whether an electron behaves like a particle or a wave. The characteristics of electrons that these experiments will reveal to me are also those of protons and neutrons, the other fundamental constituents of matter. Similar experiments are commonly performed in many physics laboratories all over the world.

In the first experiment, a beam of electrons that are going quite fast passes into air containing water vapour. Each electron causes condensation of water along its path and creates a string of tiny droplets of water marking its trajectory. The gadget, called a cloud chamber, has been widely used in the study of electrons, protons, and other basic constituents of matter. A photograph taken with such a chamber is shown in fig. III-8. The curly track at the top is from an electron, curly because of the presence of a magnetic field exerting a force on it. The other tracks are from other particles passing through the chamber. Clearly the electron is localized, and its momentum and energy can be measured in various ways. We therefore conclude that electrons are particles.

In the second experiment, a beam of electrons travelling at various speeds, and therefore with a spread of momenta and energy, bounces off the flat surface of a nickel crystal. One might expect that each electron, being a particle, would scatter off a nickel atom on the surface and proceed in some arbitrary direction; there should be no noticeable pattern in the directions of the reflected electrons. What actually happens is that all the electrons of a given energy come off in one and the same direction, and this direction is different for different energies. Just as sunlight is diffracted by the pattern on a compact disc into a spectrum of different colours (frequencies), the stream of electrons is diffracted by the regular atomic pattern of the crystal into a spectrum of different energies.

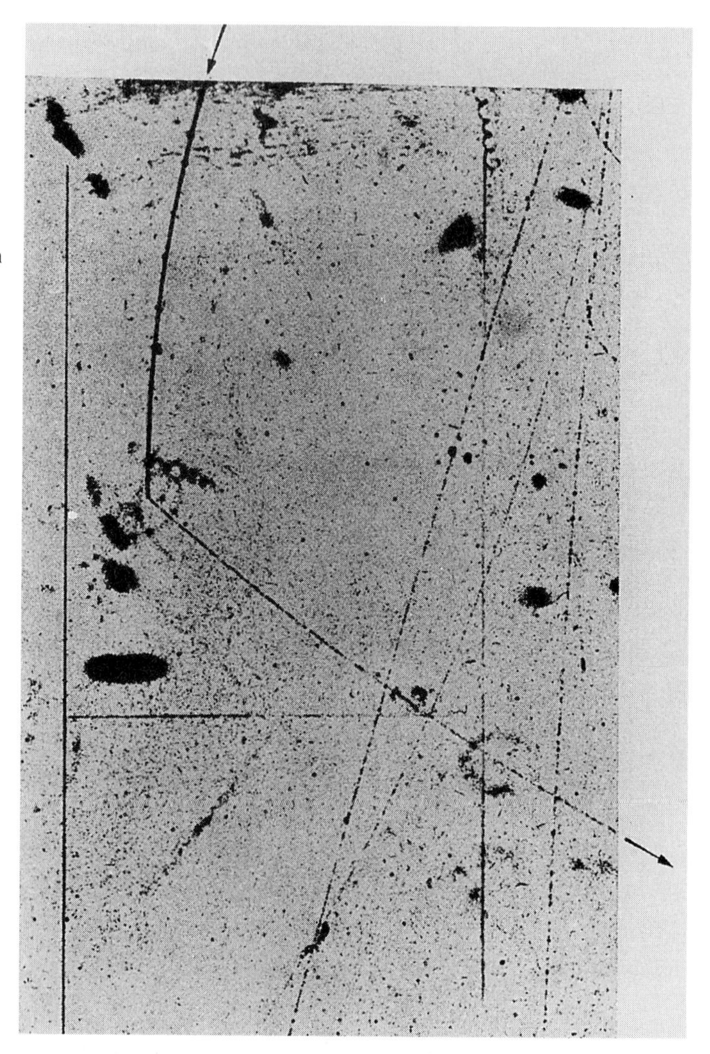

FIGURE III-8. A cloud chamber photograph. The curly track near the top is from an electron, and the other tracks are from other particles. The kinked track is from the transformation of one particle at the kink to another. The curvature of the tracks is due to a magnetic field which exerts a transverse force on moving charged particles. (Reproduced with permission from Handbuch der Physik, vol. XLV, p.275, Springer-Verlag, Heidelberg 1958.)

Table III-2 shows the one-to-one correspondence between the two cases of light diffracted by a compact disc and electrons diffracted by a nickel crystal.

Table III-2. *Comparison of the diffraction of light and electrons*

waves (light)	particles (electrons)
colour, same as frequency	energy
colour spectrum	energy spectrum
compact disc	nickel crystal
regular pattern of dots on disc	regular pattern of atoms in crystal
diffraction of light	diffraction of electrons

It appears that in the experiment with the crystal of nickel, the electrons behave not as particles, but as waves with a frequency that is somehow related to their energy, and these waves are diffracted by the regular atomic arrangement in the crystal. Such an experiment was first done by Clinton J. Davisson and Lester H. Germer, and independently and at about the same time by George P. Thomson and A. Reid.

In the third experiment, the electron waves coming off the nickel crystal pass through a cloud chamber as in the first experiment. One would expect them as waves to spread throughout the chamber, but instead the electrons behave like particles again, making tracks in the chamber just as in the first experiment. So it seems that, depending on how they are observed, electrons behave like either waves or particles. The very same electrons that behaved like waves when they bounced off the crystal of nickel seem to turn into particles as they go through the cloud chamber. This situation is totally alien to everyday experience, and we need a new way of thinking about it. This way exists, and it is called *quantum mechanics.*

The above experiments show what is called the *wave-particle duality* of electrons. I should not leave you with the impression that this is all the evidence there is for it. A vast number of other kinds of experiments have been done, and they all lead to the same conclusion. Experiments with protons, neutrons, atoms and molecules show that they also exhibit wave-particle duality and behave, depending on the experiment, like waves or like particles. There are other primary constituents of matter whose names you might have sometimes seen in the science pages of newspapers and magazines, like neutrinos and quarks. They too show wave-particle duality, and are described by quantum mechanics. This duality is a fundamental and universal property of matter. There are more surprises to come, as you will see in the next section.

8 Light: waves and particles

Light shows interference and diffraction effects. It is a form of electromagnetic wave motion. These waves come in a tremendous range of wavelengths and frequencies, of which ordinary light forms only a very small part. They all travel at the same speed, namely the speed of light. We call the waves in different ranges of frequencies by different names. Some examples, in order of increasing frequency, are radio waves, microwaves, infrared radiation, visible light, ultraviolet radiation, X-rays, gamma rays. Having seen that electrons behave sometimes like particles and other times like waves, we might wonder whether the same is true for light also. Now the simplest experiment we can do with two particles, two billiard balls say, is to shoot one at the other so that they collide and move off in different directions. We find that the total momentum and energy of the two are the same after the collision as before. A billiards player can use this result from physics to improve his game. Now suppose we take a bunch of electrons that are standing still, and shoot a beam of X-rays at it, and look at the scattered X-rays. I find that I can describe the result exactly as if the X-rays consisted of a stream of particles, whose energy and momentum are related to the frequency and wavelength respectively of the X-rays, and these particles bounce off the electrons keeping the total momentum and energy constant. I conclude that these waves, in this case, behave like particles. These particles of X-rays are called photons. Such an experiment was first done by Arthur H. Compton. I note from its result that wave-particle duality holds for electromagnetic waves also. As we shall see later, the exact relationships between energy and frequency, and momentum and wavelength, are the same for these waves as they are for electrons. Wave-particle duality is a property of all of nature.

9 Wave-particle duality

Now we face the following question: a tennis ball, say, is made up of electrons, protons and neutrons, all of which have wavelike properties. Does that mean that I must then think of the ball also as a wave that can be diffracted and so on? This should certainly make a game of tennis rather complicated, with the ball being diffracted around the racket. Fortunately for tennis, and indeed for all of our normal activities, we can forget wave-particle duality for everyday objects. A ball behaves like a particle, ripples on a lake are waves, with separate and distinctive characteristics that were listed in table III-1 (page 47). We say that we are looking at the macroscopic world in such cases. When we wish to understand the properties of matter in terms of the electrons and nuclei in it, at the so-called microscopic level, we cannot do without wave-particle duality.

It could be that some of you are having difficulty in accepting this fundamental duality of nature at the atomic level, a duality that is not evident in our everyday world. After all, things around us seem to be clearly distinguishable as either particles or waves. So how is one to picture an electron as being both a wave and a particle, exhibiting the characteristics of the one or the other but not both at the same time, depending on what one does with it? Well, the following might suggest to us that the basic idea of duality is not very far-fetched. My friend Brian does physics, and also plays the piano, but not both at the same time. He goes into the laboratory and becomes a physicist; he sits down at the piano and becomes a pianist. In each case he is still the same person called Brian, displaying one aspect or the other depending on where he finds himself. Indeed, most people have many aspects to their personae. We can think of nature as being rather simple in having only two.

The description of the world that embodies its wave-particle duality is, as I mentioned before, called *quantum mechanics.* You may in your reading elsewhere sometimes come across the term wave mechanics. It means the same as quantum mechanics, but the term is falling into disuse. I shall introduce you in the rest of this chapter to some of the basic results of quantum mechanics.

10 Some formulas of quantum mechanics

I shall be writing down some very simple formulas in this section. You can think of them as just an economical way of saying the same thing in words. I shall express in words the first formula when I get to it, so that you can see what I mean. You may if you wish do the same thing for yourself when you encounter later formulas.

We saw that we can characterize a moving particle by giving its position, momentum (equal to its mass multiplied by its velocity) and energy, and a wave by giving its amplitude, wavelength and frequency. As before, we shall use the symbols $\mathbf{p}$ for the momentum vector (and p for the magnitude of the momentum), E for energy, A for amplitude, λ for wavelength and f for frequency. These symbols stand for the numerical values of the quantities for any particular case that we are considering.

Before we proceed, I must introduce you to another way in which to think about the wavelength of a wave. Instead of saying that the wavelength is λ centimetres, we count the number of crests per centimetre of the wave (which is equal to $1/\lambda$), call it the wave number and give it the symbol k. Thus $k = 1/\lambda$ per cm, which can also be written $1/\lambda\ \text{cm}^{-1}$. The symbol cm^{-1}, read 'reciprocal centimetre', stands for 'per centimetre'. A wavelength of one-quarter centimetre is the same as a wave number of one divided by one-quarter, or four, reciprocal centimetres. The wave is also travelling in some particular direction, and we can include this information also by introducing the wave vector $\mathbf{k}$, whose size is given by k as defined above and whose direction is the direction of travel of the wave.

So now we have light and electrons, both of which behave sometimes like particles with momentum $\mathbf{p}$ and energy E, and other times like waves with wave vector $\mathbf{k}$ and frequency f. Is there any relationship between the particle-like quantities and the wave-like quantities of either light or electrons? The answer to this question has been given by innumerable experiments, and reveals a profound and wonderful simplicity and symmetry of nature that lay hidden from us till the beginning of the twentieth century. I say symmetry, because these relationships apply to electrons as well as to light waves. I first write down the answer, which consists of two simple formulas, and then explain their meaning:

$$\mathbf{p} = h\mathbf{k}$$

and

$$E = hf$$

In words: *the momentum of an electron (or proton, light wave, ...) is equal to its wave vector multiplied by a particular number that I have denoted by h, and the total energy is given by the frequency multiplied by exactly the same number.* This is a unique number that is fixed by nature, and we have no control over it whatsoever. If I express masses in grams, distances in centimetres, and time in seconds, then h has the value 6.6251×10^{-27}. You might wonder why I did not just put this number into the formulas, instead of denoting it by the letter h. The reason is not just laziness, but because the number would be different if I chose to measure masses and distances in pounds and yards respectively, say. Further, its actual numerical value may change slightly as more and more accurate ways of measuring it are used. The constant h is called the *Planck constant*, after the physicist Max Planck.

I have expressed the energy of the electron in terms of its frequency. I recall that the kinetic energy K of the electron, namely the energy due to its motion, can be expressed in terms of its momentum p and mass m:

$$K = \frac{p^2}{2m}$$

and since the momentum is related to the wave number k by

$$p = hk$$

I put this value of p in the formula for K and find that

$$K = \frac{h^2k^2}{2m}.$$

In words: the kinetic energy K is equal to the momentum p multiplied by itself (which I call *p-squared*) and the result divided by twice the mass m. Since the momentum itself is equal to the Planck constant h multiplied by the wave number k, the kinetic energy is got by multiplying h-squared by k-squared and dividing the result by twice the mass. I think you will agree with me that formulas give us a very convenient way of expressing what would take a lot of words to say!

Quantities like the Planck constant h, the mass of an electron, a proton or a neutron, the speed of light in free space (denoted by the letter c), the magnitude of the electric charge of an electron or a proton (denoted by the letter e), their intrinsic angular momenta (plural of momentum) and a measure (called the *magnetic moment*) of their behaviour as bar magnets, are all called *fundamental constants*. As far as we know, they are the same on earth as anywhere else in the universe, and they are the same now as they were at all times past and as we expect them to be in the future. We do not have much of a clue as to why they are constants, or why each has the particular value that it has. We must simply accept them as expressing some profound fundamental properties of the natural world.

11 Particle-waves: probability

Let us go back now to the electron that behaves sometimes like a particle and other times like a wave. I talk about the electron here just to have something specific in mind; the picture applies equally for any other particle. To think of the electron as a particle does not present us with any problem; we may imagine it to be like a very tiny speck of dust moving about. At any given time we can say that it is there and has that momentum. At the same time it has a wave-like nature that shows up in the right kind of experiment. This wave has a frequency f, a wavelength λ (or wave number k), and an amplitude A. We saw that the frequency and wavelength are connected with the energy and momentum through the Planck constant h. We now want to give a meaning to the amplitude. In so doing, we discover a startling new aspect of nature that is hidden from us in our everyday experience.

Let me consider an electron that is travelling in a straight line with some momentum p. Associated with this electron is a wave of wavelength equal to h/p, and an amplitude A. If I multiply the amplitude A by itself, I get the number A^2, which is always positive or zero. If I now do an experiment to try to find if the electron is in some particular part of its path, I find that it has a *probability* of being there which is proportional to the value of A^2 there! The larger the value of A^2, the more likely is it to be there; and if A^2 is zero in some region, the electron will definitely not be found there.

Let us pause a bit and see what exactly is meant by probability in this context. If a coin is tossed many times, it will come up heads in half the tosses and tails in the other half. We say then that the probability that a given toss of the coin gives heads (or tails) is a half. Of course this does not mean that there is any uncertainty about the result after the toss has been made: it will definitely be heads, or definitely tails. The idea of probability as we use it here has meaning only when applied to the results of a large number of tosses of the coin. It is in this sense that we talk about the probability associated with an electron wave. I imagine that I do a very large number of experiments one after the other, all under identical conditions, to find where the electron is. I shall then find that the electron is seen in more experiments in a region where A^2 is large than in a region where it is small. In the limit, if the number of experiments becomes infinite, then the ratio of the two values of A^2 equals the ratio of the numbers of times the electron is found in the two regions. If the amplitude is zero in some region, then I shall never find the electron there.

With this interpretation of the amplitude A of an electron wave, first proposed by the physicist Max Born, we have left the world of everyday experience and entered the quantum world. To illustrate the difference between the two, let me take an example from each world: a train that I know to be somewhere between two points P and Q, and an electron that I also know to be somewhere between those two points. I know that I can find out exactly where the train is at any given time, if I want to; probability just does not enter the picture. Things are quite different with the electron. Since I know definitely that the electron is not outside the region between P and Q, its wave in this outside region must have zero amplitude. Inside the region the electron has a wave associated with it, whose precise shape we shall come to in a moment, and whose amplitude A varies as we go from P to Q. We cannot even in principle talk about the electron definitely being in some part of the region between P and Q, but only of the probability of finding it there, a probability proportional to the value of A^2 there.

Quantum mechanics thus tells us that we can only determine the probability of the result of an experiment to locate the position of an electron. This is a profound break with the way we have been used to thinking about the natural world around us. The statement 'the electron is definitely there' is replaced by 'there is a probability that I shall find the electron there if I look for it'. You will note that this probability refers to an actual effort, a set of experiments, to find out where the electron is.

Some of you might begin to wonder what this quantum-mechanical way of looking at things does to our intuitive ideas of the reality of the world around us. What we were used to thinking of as certainties, we find we should now think of as probabilities. Since the advent of quantum mechanics in the early part of the twentieth century, such questions have engaged many philosophers of science, and some physicists too, Albert Einstein and Nils Bohr among them. Most physicists, however, take a somewhat pragmatic stand on such questions. Their attitude is roughly the following: quantum mechanics is a description of nature that works correctly everywhere it is applicable, and leads to predictions that are subsequently verified; and that is all that one can ask of a theory in physics. This attitude is in tune with my purpose in writing this book, namely simply to introduce you to the quantum-mechanical description of the properties of matter and to resist the temptation to dwell on philosophical implications.

The quantum-mechanical interpretation of the square of the amplitude as a probability immediately leads to a dramatic new effect that is the justification for the name quantum mechanics. I describe this effect in the next section.

12 Particle-waves: the quantization of energy; the wave function

Suppose I want to find out a bit more about an electron as it is moving: what its energy is, for example. If all I know about its location is that it is somewhere, then I would not have an easy time finding it. It could be anywhere in the universe. Common sense tells me that I would do better to trap it in a box from which it cannot get out, so that I know that it is definitely somewhere inside the box, although I can only talk about the probability that it is in any particular region in the box. If now I can find out what the wave is which is associated with this electron, then I can hope to learn something about its motion. I remind you that there are relations (p. 51) between wave properties like frequency and wavelength on the one hand, and particle properties like energy and momentum on the other.

I take a very simple example that leads to a startling new result, namely *the quantization of energy*. I mean by this that *the energy can only have any one of a set of discrete values*, but not anything that lies between them. For my example, I imagine that I have trapped an electron between two parallel walls P and Q which it cannot penetrate, and that I have further arranged matters so that it can only move back and forth along a straight line and cannot stray from this line (fig. III-9). L is the distance between the walls. I summarize all this by saying that the electron is trapped inside a one-dimensional box of length L.

FIGURE III-9. A one-dimensional box of length L extending from P to Q. An electron is trapped inside the box, and all we know is that it is somewhere between P and Q.

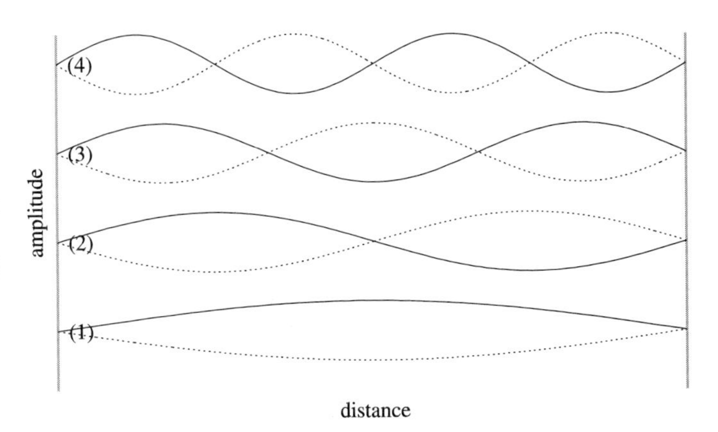

FIGURE III-10. Four of the possible standing waves which can represent an electron trapped in a one-dimensional box. The wave labelled (1) has the longest wavelength, (2) the next longest, and so on. Note that no in-between wavelengths are possible. Each wave is to be imagined as oscillating between the continuous and broken lines depicting it.

If I now look to see where the electron is at any given instant of time, one thing is certain: I shall not find it outside the box. If there is no electron, there is no wave. So the amplitude of the electron-wave outside the box is zero, and it remains zero right up to the two walls.

Since the electron is definitely somewhere inside the box, its amplitude inside cannot be zero. However, its amplitude on the two walls must be zero, the same as outside the box, because the amplitude of a wave cannot change abruptly. I now try to draw a picture of such a wave, and I find that there is more than one possibility, depending on how many crests (and troughs) of the wave are fitted inside the length L. Further, I find that there is a maximum possible wavelength which can be fitted inside the box, corresponding to just one crest (or trough). Other successive possibilities with decreasing wavelengths would be one crest and one trough, two crests and one trough, two crests and two troughs, …, all fitting inside the box. Figure III-10 illustrates these waves. You must imagine that each wave is vibrating between the positions shown by the continuous and broken lines somewhat like the vibrations of a plucked string in a guitar. There is an important point to note here: the wavelengths can only have those values which permit integer numbers (1, 2, 3, …) of troughs and crests to fit inside the box.

At this point, you may have noticed that the waves I have just illustrated do not look quite like the waves that we encountered earlier. Those earlier waves were travelling somewhere with some velocity, while these waves are trapped inside the box, and are not going anywhere. Such a trapped wave is called a *standing wave*, to distinguish it from the other sort which is called a *travelling wave*. A standing wave results from adding together two travelling waves of the same frequency and wavelength, which are travelling in opposite directions. This is illustrated in figs. III-11 and III-12. The two waves can be thought of as arising from reflections at the ends P and Q of the box in fig. III-9. An everyday example of a standing wave is the way the strings in a musical instrument like a piano or a guitar vibrate.

Thus we see that I cannot fit a wave of any arbitrary wavelength inside the box; it can only have certain wavelengths that are fixed by the length of the box, and none

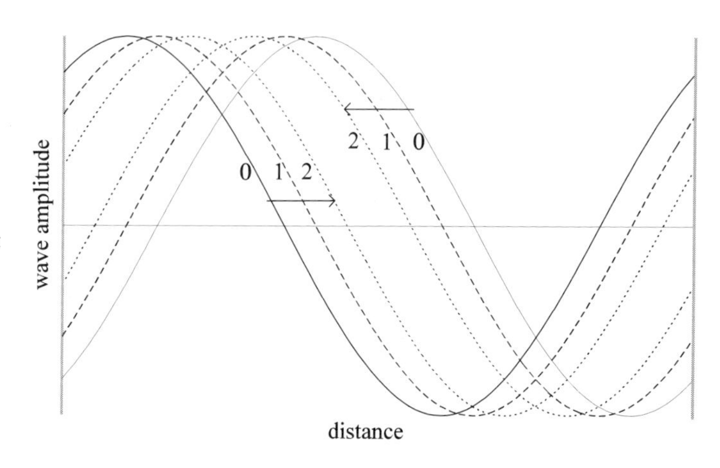

FIGURE III-11. A wave travelling to the right in a one-dimensional box, and the reflected wave travelling to the left, as indicated by arrows. Each wave is shown at the same three successive instants of time labelled 0, 1, and 2. The wavelength is such that the two added together result in a standing wave, as in fig. III-12.

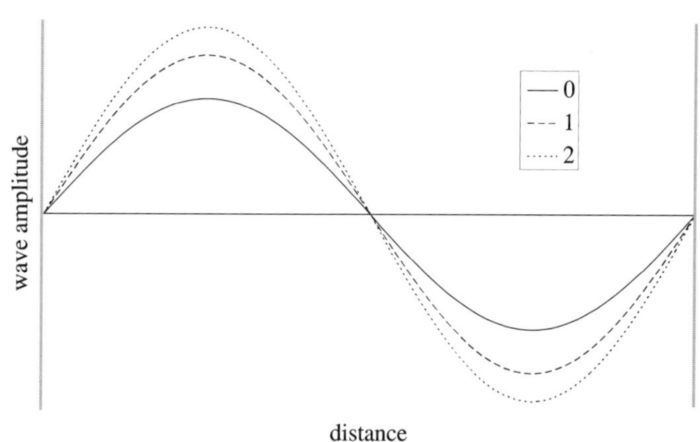

FIGURE III-12. What the standing wave, produced by adding together the two travelling waves of fig. III-11, looks like at the same three instants of time 0, 1, and 2

that lies in between them. We recall that the wavelength, or equivalently the wave number k, is related to the energy E of the electron by the formula

$$E = \frac{\hbar^2 k^2}{2m}.$$

So we conclude that the energy of the electron can take only certain discrete values, and none in between. In such a case, we say that the energy is *quantized.* We refer to these values of the energy as the *energy levels* of the electron in the box. This result is entirely due to the wave-particle duality of the electron. If the electron were to be considered purely as a particle in the box, then it could have any desired value of the momentum and energy; there would be no quantization. The smallest energy it could have would be zero, when the particle is just sitting motionless somewhere in the box. With the quantization of energy, we have arrived at a fundamental result of quantum mechanics, which sets it apart from our everyday understanding of the world around us. The electron cannot sit still; it is forced to have a minimum energy, corresponding to the longest possible wavelength, and it can also have only a set of discrete higher energies. The energy is quantized.

We shall now do a simple calculation to get a formula for these allowed energies, the energy levels. The need to fit a whole number (1, 2, 3, ...) of troughs and crests into the box means that the wavelengths cannot be made to be anything we wish, but can only have certain values that are fixed by the length L. The wavelength is the distance occupied by a crest and its neighbouring trough. Now look at the longest wave, labelled (1), in fig. III-10. There is just one crest (which with time alternates with a trough because of the vibrations of the wave), and its length is L, which is also the length of its trough. So its wavelength is equal to twice L, or $2L$. Its wave number, which is just the reciprocal of its wavelength, is then $1/(2L)$. This is the smallest wave number that the electron-wave can have in the box, and I shall denote it by the symbol k_1:

$$k_1 = \frac{1}{2L}$$

where the subscript 1 in k_1 reminds me that I am thinking of the wave number of the wave labelled (1) in fig. III-10. I now come to the wave labelled (2). I see from the figure that its wavelength is exactly equal to L, and therefore its wave number, which I shall call k_2, is equal to $1/L$, or

$$k_2 = \frac{1}{L} = 2\,\frac{1}{2L} = 2\,k_1.$$

For the wave labelled (3), I see from the figure that there are $1\frac{1}{2}$ wavelengths in the distance L, so that its wavelength is $\frac{2}{3}L$, and its wave number k_3 is

$$k_3 = 1/\tfrac{2}{3}L = 3\,\frac{1}{2L} = 3\,k_1.$$

QUESTION What is the wave number k_4 of the wave labelled (4)?

ANSWER There are exactly two wavelengths in the length L, so that the wavelength is $L/2$, and so the wave number k_4 is $2/L$, which I can write as

$$k_4 = 4k_1.$$

I see a pattern beginning to emerge: the wave numbers of the successive waves which fit inside the length L are k_1, $2k_1$, $3k_1$, $4k_1$, ... For the n-th wave, where n is an integer which has any one of the values 1, 2, 3, 4, ..., the wave number, which I shall denote by k_n, will be n times k_1:

$$k_n = nk_1, \text{ where } k_1 = \frac{1}{2L}\,.$$

I have introduced in the foregoing a new symbol, k_n. I read it as 'k sub n', to remind me that n is written a little bit below k, i.e. n is a subscript. If $n = 1$, then I read it as 'k sub 1', and so on. You can see that with the one formula that I have just written for k_n, I do not need a separate formula for the wave number of each wave. I now have a useful shorthand when I wish to make a statement involving the wave number, which applies to all the waves. For example, suppose I want to find the energy of each of these electron-waves. I assume that there is no potential energy, so that the total energy is just equal to the kinetic energy, which is equal to

$$\frac{h^2k^2}{2m}$$

for an electron of mass m and wave number k. Let E_n be the energy of the n-th wave with wave number k_n. Then we have

$$E_n = \frac{h^2(k_n)^2}{2m}.$$

We saw above that k_n is just n times k_1, and so we can write

$$E_n = \frac{h^2(nk_1)^2}{2m}.$$

Now, using the value for k_1 of $1/(2L)$, we get

$$E_n = \frac{h^2n^2}{8mL^2}$$

We get the energy E_1 of the wave with the longest wavelength by putting n equal to 1:

$$E_1 = \frac{h^2}{8mL^2}$$

and we can now write the energy of the n-th wave as

$$E_n = E_1 n^2.$$

QUESTION We said that the matter waves are also vibrating. What is the frequency f_n of the wave whose energy is E_n?

ANSWER Since the energy is given by the Planck constant h multiplied by the frequency, the frequency f_n equals E_n divided by h.

Let me state in words our result:

The energy of an electron in a one-dimensional box is quantized. The smallest energy it can have, denoted by E_1, cannot be zero but has a finite value that becomes larger as the length of the box decreases. The other values of energy that it can have, namely its energy levels, are given by multiplying E_1 by n^2, where n is given one of the values 2, 3, 4, ... These are the only possible values of energy that the electron can have, and they belong to the allowed quantum states of the electron.

The lowest energy that the particle can have is called its *zero-point energy*, and is a direct consequence of its wave nature. If it were simply a particle, its lowest energy would be zero and it would be sitting motionless. This however is not allowed because of its wave nature: instead, it has a zero-point motion which results in its zero-point energy.

So much for the energy of the electron in a one-dimensional box. How about its momentum? We saw that the momentum vector **p** is given by

$$\mathbf{p} = h\mathbf{k},$$

where **k** is the wave vector. In the present case, there are two waves with the same wavelength that are travelling in opposite directions. The corresponding wave vectors are therefore $+\mathbf{k}$ and $-\mathbf{k}$ respectively, and the probable momentum that we would measure is zero.

I have taken the example of an electron in a one-dimensional box, to keep things simple while bringing out the important consequences of quantum mechanics. The world around us, however, is three-dimensional: we shall later, for example, want to talk about electrons in a real crystal. It turns out that the basic features of the quantum description still hold. The energies of the electrons can have only discrete values that are determined by the size and nature of the crystal. Also, the dependence of energy on the wave number becomes more complicated than for the simple case that we have considered above. This is because the electric potential energy of the electrons, which have an electric charge, must also be taken into account. The physicist uses a mathematical equation called the *Schroedinger equation,* named for Erwin Schroedinger who first wrote it down, in such cases. If I know the conditions under which the electron is moving, for example that it is moving back and forth inside a one-dimensional box of length L, then I can write down the Schroedinger equation describing its motion, and from it I can calculate the amplitudes and energies corresponding to the different possible waves associated with the motion. What I have done above with words, this method allows me to do with mathematics. The Schroedinger equation also permits me to calculate the electron's energies in more complicated cases, such as when it is in a hydrogen atom or in a crystal. We shall not do such calculations in this book, but shall look at some of the results in later chapters, and see that they are what we would have expected from the understanding of quantum mechanics that we are now acquiring here.

I said that the square of the amplitude, A^2, at any point is proportional to the probability of finding the electron there. I show in fig. III-13 what this quantity looks like for the four waves labelled (1) to (4) in fig. III-10. If I were to think of the electron as simply a particle, I would say that it has the same probability of being anywhere inside the box. When I use the quantum description of the electron, I find by looking at the figure that the electron is more likely to be found in some parts of the box than in others. For example, for the wave labelled (1), the electron is most likely to be found at the centre of the box, and progressively less and less likely as I move away from the centre in either direction. It will definitely not be found at the two ends of the box or outside it,

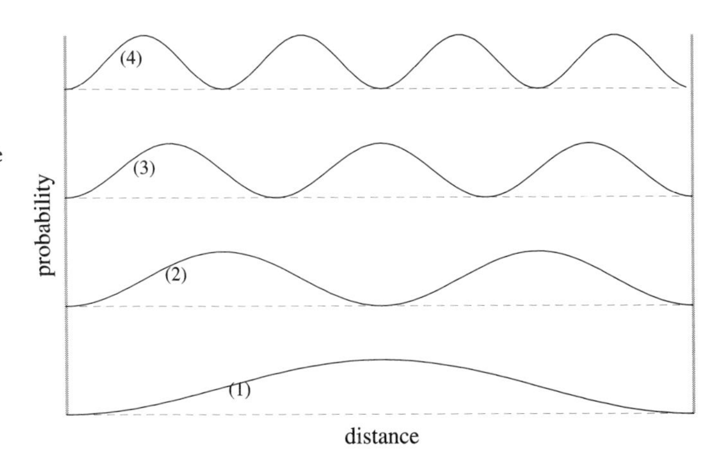

FIGURE III-13. The square of the amplitude at a point of the wave describing a particle gives the relative probability of finding the particle in the vicinity of the point. This probability is shown here, labelled the same way, for each of the waves which describe an electron in a one-dimensional box (fig. III-10). Note that the probability is not at all constant everywhere, but varies between a maximum and zero.

where A^2 is zero. So we have here yet another example of how the quantum world is totally different from the everyday world. I should remind you of the meaning of probability here: it refers to the different outcomes of a very large number of experiments that one imagines to be done to see where the electron is in the box.

Some of you may have begun to wonder what these seemingly strange ideas have to do with our usual sense of reality. Let me assure you that innumerable experiments have shown that quantum mechanics is the correct description of nature. It provides the answers to the kind of questions that were posed at the beginning of chapter I, as you will see in later chapters. We must simply accept that this is the way nature is at the microscopic level of atoms and electrons and nuclei, strange as it may seem compared to what our senses tell us about the macroscopic world around us.

We saw that an electron confined in an enclosure is associated with a wave that has some amplitude at every point inside the enclosure. I can label each of these points by using vectors, which you recall are things that have a magnitude and a direction. I first fix one point, which I call O (for origin). Then I take the point P which I want to label, and imagine a straight arrow whose tail is at O and tip at P. This arrow has a magnitude, namely its length, and a direction, from O to P. So I can represent the arrow by a vector; I call it **r**, the position vector of P with respect to O. By changing the length and/or the direction of the arrow, which is the same as saying that the vector **r** is a variable, I can cover all points in the enclosure. I use the Greek letter ψ (pronounced *psi*) as the symbol for the amplitude of the wave, and write the amplitude at the point whose position vector is **r** as $\psi(\mathbf{r})$. This way of writing is to remind me that ψ is a function of **r**, i.e. the value of ψ can be different at different points in space. I call $\psi(\mathbf{r})$ the wave function. This single symbol $\psi(\mathbf{r})$ stands for the amplitude of the wave at every point in the enclosure holding the electron. I think you can appreciate the usefulness and economy of this notation. The Schroedinger equation, which I mentioned earlier, allows me to determine the wave function and the energies of the electron. I must also keep in mind that the wave function is oscillating in time at each point in space, the frequency being equal to the energy divided by the Planck constant h.

For an electron in a one-dimensional box different waves are possible, and we labelled them n, where n takes on one of the values 1, 2, 3, 4, ... Each of these waves can be described by a wave function. We can write the wave function for a given value of n as $\psi_n(x)$, where x is the distance measured from one end of the box. So the value of x can vary from zero to L where L is the length of the box. The corresponding energy level will then be E_n. The number n, which first appeared in the formulas for the wave number and energy (p. 57) and with which I have now labelled the different wave functions and energy levels, is called a *quantum number*. As each value of n corresponds to a particular value of the wave number k_n, I can equally well use the k_n as the quantum numbers to specify the different wave functions. Quantum numbers are a characteristic feature of quantum mechanics, and we shall see other examples in later chapters. We see now that the waves shown in fig. III-10 are just pictures of the wave functions for the first four quantum states of an electron in a one-dimensional box.

Now some mathematics for those of you who remember trigonometry. The wave functions for the electron in a one-dimensional box are

$$\psi_n(x) = A \sin\left(n \frac{\pi x}{L}\right), \text{ with } n = 1, 2, 3, 4, \ldots,$$

as you can verify from fig. III-10. If this formula looks opaque to you, just skip it and carry on!

13 Summary

This chapter contains a number of new ideas and ways of looking at things. I shall give you now a summary of the principal points in the chapter, section by section. It may be helpful for you to look up the relevant sections as you go through this summary.

I look in sections 1 and 2 at the world around us, and note that everything is in a state of *motion*. I find that I can put the things that are moving into one of two classes: they are either *particles*, or they are *waves*. A particle is an idealization of our everyday sense of a moving object: it has mass and occupies a tiny volume. We think that electrons and nuclei might be good examples of particles, because they have a mass and they are certainly very small. We shall find later that this is not the whole story. The idea of a wave is easy to grasp: we have all seen waves on the surface of water.

In section 3 I look in more detail at the motion of a particle. Besides its position and mass, the moving particle has *velocity*, *momentum*, and *energy*. The energy comes in two forms, *kinetic energy* which is due to the motion, and *potential energy* which requires some special trick to convert it into kinetic energy and thus to motion. I introduce the idea of a *vector*, which is a quantity that has both a *magnitude* and a *direction*, and denote it by a symbol in heavy type. I call a quantity that has only a magnitude, but no direction associated with it, a *scalar*. Momentum and velocity are examples of vectors, whereas mass and energy are examples of scalars.

Section 4 is a brief diversion from the main course, to introduce you to the *conservation of momentum and energy*. These conservation laws form the fundamental basis for all of physics. They are a consequence of the nature of space and time.

I describe the characteristics of waves in section 5. A wave has *amplitude*, *wavelength*, and *frequency*. Unlike a particle which is localized in space, a wave is spread over space. It is travelling in some direction with some velocity. A wave is *diffracted* around an obstacle in its path, and continues onwards. Two waves that come together *interfere* with each other and produce a new wave.

In section 6, I recapitulate the characteristics of particles and waves, and present a table that emphasizes the distinction between them. It appears that in the world we see around us, motion can be unambiguously assigned either to particles or to waves, with no overlap between the two.

This picture is dramatically changed when we look at the world on the scale of an electron (or proton, neutron, etc.). Depending on the kind of experiment I do on it, an electron may move like a particle or like a wave, with the corresponding features shown in table III-1. This wave-particle duality is the subject of section 7.

We now come to section 8, and find that nature has not yet finished with its surprises for us. Light as well as other forms of wave motion shows particle-like or wave-like properties depending on the specific way in which they are observed. The particles of light are called *photons*. Wave-particle duality thus turns out to be a property of all nature.

Section 9 contains some general remarks on this duality, and points out that we can still continue to think of particles and waves as distinct entities on the scale of the macroscopic world.

The quantitative description of wave-particle duality is called *quantum mechanics*. I introduce you in section 10 to the basic formulas of quantum mechanics relating the wave-like and particle-like properties of the electron (or proton, neutron, etc.) to each other. Exactly the same formulas hold between the wave-like and particle-like properties of light. You may think of this as a manifestation of some deep symmetry in nature.

I continue the account of quantum mechanics in section 11, and describe how the amplitude of the wave is related to the probability of finding the particle there. The amplitude multiplied by itself, which I call the square of the amplitude, is proportional to this probability.

I illustrate in section 12 this interpretation of the amplitude by looking at an electron trapped in a one-dimensional box. I find that the electron can only have any one of a set of values for its energy, and nothing in between. Furthermore, its lowest possible energy cannot be zero. This means that the electron cannot be sitting still, but must necessarily be moving with at least this minimum energy. This is one more example of the unexpected consequences of wave-particle duality. I conclude the section by describing what is meant by the wave function for the electron.

This brings me to the end of my summary. In the chapters to follow I use quantum mechanics to describe a crystal, and see how I can then understand its properties. A crystal is composed of a very large number of atoms. We have so far looked at the quantum mechanics of an electron in a box, but this is very different from an atom, let alone a whole crystal. The simplest atom we have is the hydrogen atom, with just one proton and one electron. I shall show you in the next chapter what quantum mechanics has to say about the hydrogen atom. The resulting picture, with some modifications, applies to other atoms also. We shall then be closer to being able to describe a crystal in quantum mechanics.

IV THE ATOM

1 Introduction

I give in this chapter the quantum-mechanical description of an atom. I look in some detail at the hydrogen atom, which is the simplest atom we have, with just one proton and one electron. From the picture that emerges, we shall be able to see how to describe all other atoms. Before continuing, I need to explain what I mean when I say that the proton and electron have equal and opposite electric charges. It is because of these charges that the atom is stable, and does not spontaneously break up into the two separate particles.

2 Electric charge and potential energy

If I take a ball in my hand and then let go, it drops to the floor. This is because of the gravitational interaction between the ball and the earth. The ball had gravitational potential energy when I held it, and when I release it, this energy is progressively converted to kinetic energy as it falls. Gravitation makes two bodies move towards each other, and not away from each other: the gravitational force is always attractive.

As a result of many experiments done over the years, we find that there is another kind of force in nature, somewhat analogous to, but in one aspect different from, gravitation. I take two idealized examples, and use them to show the features of this force. First I consider two electrons that I imagine to be pinned down a distance r apart. If I now unpin the electrons so that they are free to move, I find that they fly apart, moving away from each other. There appears to be a repulsive force between them, which therefore cannot be due to the gravitation (which attracts them to each other) between the two masses. Just as the source of gravitation is the mass of the electron, there must be a source for this new force: I call it *the electric charge*, or simply the charge, of the electron.

I should now like to make a small detour, and say a few words that will give you an insight into some of the ways of physics. Having been told that an electron has something called charge, you might be tempted to ask, 'Well, what exactly is it? Is it something that makes the electron glitter, or colours it blue, or something like that?' The answer is, none of the above. It is just a compact way of saying that the electron has a property that makes it exert a force, which has nothing to do with its mass, on another electron. The reasoning goes as follows:

1. We observe that two electrons repel each other. This is a given fact of nature.
2. This repulsion cannot be due to the force of gravitation between the electrons, because then they would be attracted towards each other.

3. So there must be another property of the electron, besides its mass, which is responsible for this force. I call this property its charge.
4. I need to be able to measure the charge – remember, physics is a quantitative science. So I need a unit and a method of measurement. I think of the analogy with measuring mass, where I have a set of standard weights labelled 1 gram, 5 grams and so on, and I find the mass of a bag of potatoes, say, by comparing the gravitational force on the bag with the force on these weights. This last step is essentially what a spring balance does. So with charge: I can make the charge on the electron my unit of charge. A superb unit it is too, for it is exactly the same for all electrons, and never changes with time. Then I can measure any other charge by finding the electric force on it due to an electron. In practice, such measurements are made by more indirect methods, but the underlying principle is as I have described it.

To conclude: when you think of electric charge, you must understand that it is just a name given to that property of the electron which is responsible for the electric force. This brings me to the end of my detour.

From the discussion of energy in chapter III, I know that the two electrons had potential energy when they were pinned a distance r apart, and this energy is converted to kinetic energy when they are released. Further experiments show me that the larger the distance r at which the two electrons are pinned, the smaller the potential energy that they have.

Now, I repeat all these experiments after replacing the two electrons by two protons. I find that I get exactly the same results for the potential and kinetic energies as before. Can I conclude that the electron and the proton must have exactly the same charge? Not yet! I do a third set of experiments, this time with one electron and one proton. Almost everything is as with the first two sets of experiments, but with one important difference: when set free, the electron and proton move towards each other and not away from each other. The force is attractive, and not repulsive as it was in the two previous experiments. It thus appears that the charge on the proton is similar to, yet essentially different from, the charge on the electron. Similar because two protons and two electrons have the same magnitude of force when they are the same distance apart. Different because two protons (or two electrons) repel each other, whereas an electron and a proton attract each other.

I can summarize all this by saying that a proton has a positive charge $+e$ and an electron has a negative charge $-e$, where e is 4.80×10^{-10} (using the cm, gram, and second as the units of measurement), and that two like charges (i.e. both positive or both negative) repel each other and two unlike charges attract each other. We have also talked earlier about another particle called the neutron. It has almost exactly the same mass as the proton, but no charge at all. Neither the electron nor the proton exerts any electric force on the neutron. I say that the neutron is electrically neutral, it has zero charge.

If you take two objects and stick them together, you get the mass of the resulting object by adding the masses of the two. It is the same with charges; only, you have to be careful with the plus and minus charges. For example, the total charge on a hydrogen atom is got by adding the charge of the proton, $+e$, to the charge of the electron, $-e$, resulting in a total charge of zero. The hydrogen atom is neutral. This electric neutrality is a property of all atoms in nature. The total negative charge of the electrons in any atom is exactly balanced by an equal positive charge of an equal number of protons in the nucleus of the atom. The nucleus has some neutrons too, but these have no charge. This neutrality of atoms is why we do not usually see electric forces in everyday life. There are some occasions, however, when some electrons are separated from their atoms, and then these forces appear. This is what happens when you sometimes find your hair crackling and standing on end when you comb it. The friction between the comb and the hair causes some electrons to detach themselves from the hair and go to the comb. So the hair is left with some ionized atoms, i.e. atoms that have lost some electrons and are therefore positively charged. Neighbouring hairs, both positively charged, repel each other, and so the hairs fan out and stand up. Some of the separated electrons jump back to where they came from, producing sparks that cause the crackling you hear.

QUESTION If an atom of carbon loses an electron, what is the charge on the resulting ion?

ANSWER Initially the atom had six protons in the nucleus with a charge $+6e$ (and also some neutrons with no charge) and six electrons with total charge $-6e$ and was therefore neutral. The ion has five electrons with charge $-5e$ and so its net charge is $+6e$ minus $5e$, or just $+e$.

We saw that two electrons held a distance r apart have electric potential energy, which is progressively converted to their kinetic energy of motion when they are let go. There is a formula for calculating the potential energy of two charges q_1 and q_2, say, held at a distance r apart. If the two charges are an electron and a proton respectively, then q_1 is $-e$ and q_2 is $+e$. The energy is given by multiplying the two charges (remembering the plus and minus signs!), and dividing the result by r. If I call this energy V, I can write for short

$$V = \frac{q_1 q_2}{r}.$$

Therefore the potential energy of the electron and the proton held a distance r apart is $-e$ times $+e$ divided by r, which is equal to $-\frac{e^2}{r}$. You remember that a negative number multiplied by a positive number gives a negative number.

QUESTION What is the potential energy of two electrons at a distance r apart?

ANSWER I multiply the two charges together, and divide by r. The answer is $-e$ multiplied by $-e$ and divided by r, or $+\frac{e^2}{r}$.

Two electrons, or two protons, have a positive potential energy and repel each other. An electron and a proton have negative potential energy and attract each other. This is an example of the connection, which holds generally, between the plus or minus sign of the electric potential energy and the repulsion or attraction.

The electric charge $-e$ is the same for all electrons, everywhere and always. The same is true for the charge $+e$ of the proton. The charge is an intrinsic and unchanging property of the particle, just like its mass. Any electric charge that we might meet in nature, since it must be made up of electrons and protons, can only have a value that is an integer multiplied by e, and nothing in between. I can have a charge of 5 times e, or 6 times e, but not, say, 5.5 times e. I can say that electric charge is quantized, and the quantum of charge is e. I should mention here that protons and neutrons are believed to be made up of particles called *quarks*, which have positive or negative charges of magnitudes $\frac{1}{3}e$ or $\frac{2}{3}e$. This internal structure of protons and neutrons has no effect on those properties of matter which are discussed in this book, and can therefore be ignored.

3 The hydrogen atom: why is it stable?

The hydrogen atom consists of a positively charged proton and a negatively charged electron, and we have seen that these two charges, being opposite in sign (plus and minus), will attract each other. So you might well ask why the electron is not sucked into the proton so that one ends up with a particle with zero charge, something like a neutron. Wc know that hydrogen atoms are quite stable in nature. Why is this so? The answer lies in the wave nature of the electron. Consider what would happen if the electron did fall into the proton and become trapped inside it. We know from many experiments that the proton is like a tiny sphere with a radius of about 10^{-13} centimetre. We recall from the last chapter (page 58) that when an electron is trapped inside a box, it must have a minimum kinetic energy that cannot be zero, and which becomes larger the smaller the box is. Well, if I try to trap the electron in a box that is as small as the proton, it will have a kinetic energy that is so great that it would come shooting out of the proton at once! When it is outside the proton, it still feels the attractive force due to the proton, and therefore cannot wander off to a great distance away. It ends up by staying somewhere near the proton all the time, and the result is the stable hydrogen atom.

Many of you are familiar with a picture of the atom that often appears in the popular press, and on banners at demonstrations about nuclear weapons, nuclear power stations, and so on. This picture looks somewhat like fig. IV-1, with the dark circle in the centre representing the nucleus, and electrons going around it in elliptical orbits. Unfortunately, this picture does not represent reality as we know it, because it does not take account of the wave nature of the electron. Quantum mechanics tells us that an electron does not go round in a fixed orbit like a planet round the sun, but may be found anywhere near the nucleus with a probability which varies from point to point.

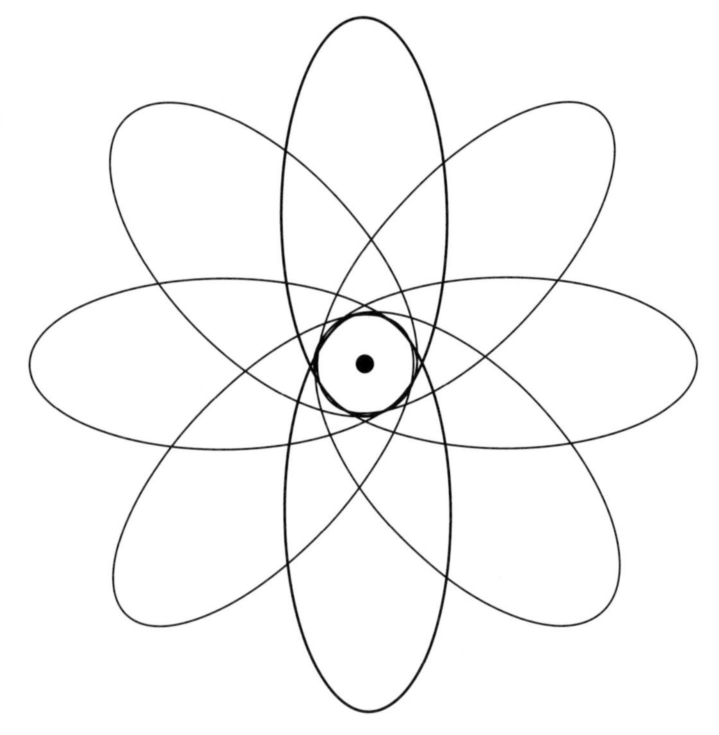

FIGURE IV-1. A popular but misleading representation of an atom, as a nucleus surrounded by electrons going around it in various orbits. The correct picture, as given by quantum mechanics, is indicated in fig. IV-3.

We have to find out what the wave function of the electron looks like when it is near the proton. Then we can find out how the probability varies from point to point and what the energy of the atom is. This we now proceed to do.

4 The hydrogen atom: the ground state

We saw in chapter III that an electron that is confined to a box has a lowest energy state, which is the one with just half a wavelength fitting inside the box. Similarly the hydrogen atom, with the electron confined to being somewhere near the proton, has a state of lowest energy. This is a general feature of any quantum-mechanical system. We call this state of lowest energy the *ground state* of the system. The physicist finds the properties of the ground state, and of the higher energy states, by solving the Schroedinger equation for the atom. We are not going to do it here. Instead, I shall show you how we can use the ideas about wave functions and probability that we learned in chapter III to get a qualitative picture of the ground state.

Look at fig. IV-2, depicting a hydrogen atom. The dark blob at the centre represents the proton, and the electron has a probability of being anywhere in the grey region surrounding it. The region extends in principle out to infinity, although the probability of finding the electron rapidly decreases as one moves away from the proton. Any point P in this region can be specified by the vector **r** connecting it to the proton. Each

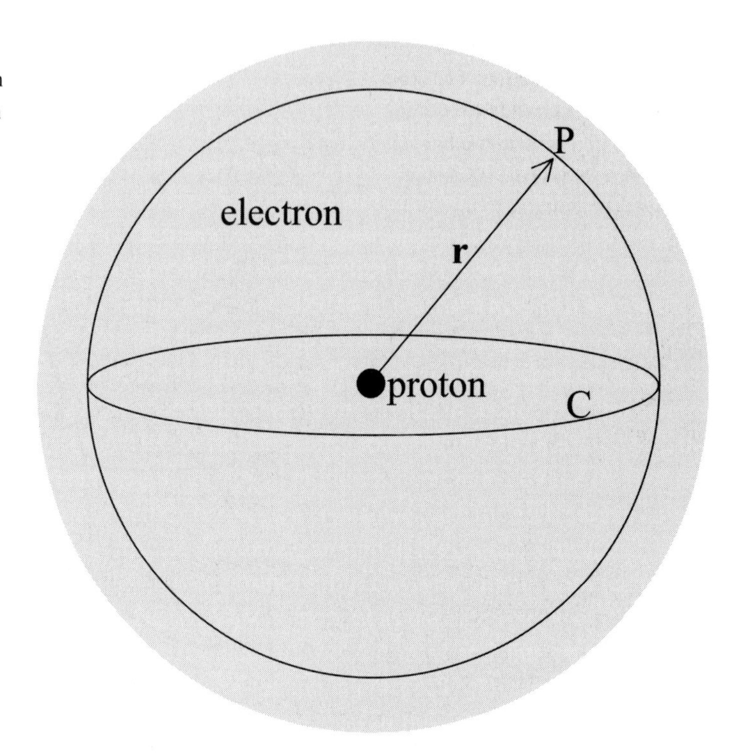

FIGURE IV-2. Illustrating a way to specify the position of an electron in relation to the proton in a hydrogen atom. P is a point whose position vector is **r**, lying somewhere on a sphere of radius *r* centred on the proton. C is a circle girdling the sphere.

quantum state of the electron has its own distinctive wave function, with a particular value at each point in the region around the proton.

Imagine now a sphere of radius *r*, which is the magnitude of the vector **r**, centred on the proton. The simplest form that the electron's wave can have as I follow it on the surface of the sphere is to have exactly the same amplitude at every point. Each of these points is at the same distance *r* from the proton. How would I expect this amplitude to change, if at all, when I go to a sphere of larger radius around the proton? Since the electron feels a force of attraction towards the proton, it is more likely to be nearer the proton than further from it. Translating this probability to amplitude, we would guess that the amplitude of the electron's wave function decreases as its distance from the proton increases. This guess is confirmed by a calculation using the Schroedinger equation.

The result is shown in fig. IV-3, where you see the probability of finding the electron in any one direction at different distances from the proton in the state described above. The height of each bar is proportional to this probability at the corresponding distance. This distance is measured in a unit of length called an *angstrom*, one angstrom being equal to a hundred-millionth of a centimetre. You can see that the electron is most likely to be very close to the proton, within a distance of about one angstrom from it. It will almost never be found at much greater distances. One can roughly say that the size of the hydrogen atom in the ground state is about one angstrom. This result, which is a prediction of quantum mechanics, is borne out by many experiments. I remind you

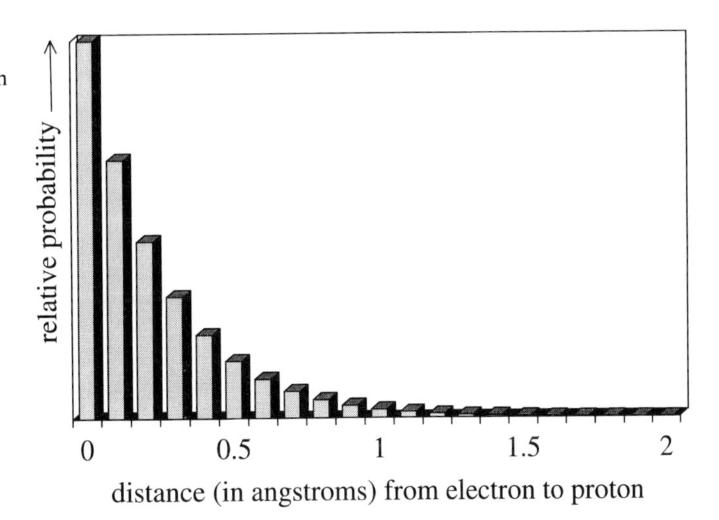

FIGURE IV-3. The relative probabilities of finding the electron in any one direction at different distances from the proton in a hydrogen atom in the ground state. The horizontal axis is divided into ranges lying within 0 to 0.1 angstrom, 0.1 to 0.2 angstrom, … . For example, the segment labelled 0 is the range from 0 to 0.1 angstrom.

again of the meaning we attach to probability in quantum mechanics. Suppose I take a hydrogen atom and do many experiments on it to find how far the electron is from the proton. The heights of the relevant bars in fig. IV-3 tell me that for every experiment in which the electron is in some direction 0.6 to 0.7 angstrom away from the proton, there will be 10 experiments in which it is in that direction less than 0.1 angstrom away.

So, instead of the popular picture of the electron going around the proton in an orbit, we now have the quantum-mechanical picture of the electron smeared out, so to speak, in a probability cloud extending out to a distance of about an angstrom from the proton. The energy of the proton with this electronic cloud is the ground state energy of the atom. Using the Schroedinger equation, one calculates this energy to be about –13.6 electron-volts. A word is in order here about why the energy is negative. I have assumed that when the electron and proton are very far apart, their energy is zero. The negative sign then means that the energy of the ground state is lower than the energy of the two particles when very far apart. If now I want to break up the atom by taking the electron very far away from the proton, I must feed an energy of 13.6 eV to the atom, because +13.6 added to –13.6 gives me zero. This is an application of the principle of conservation of energy.

I have introduced here a new unit with which to measure energy, the *electron-volt.* You are familiar with the fact that the electricity in your home is supplied at about 230 volts, or at about 115 volts if the home is in north America. You can think of volts as the pressure difference that drives the electrons to flow in a current, for example through a light bulb. The electron-volt is the energy that an electron acquires when it moves under a pressure difference of one volt. The electron-volt, abbreviated to eV, and the angstrom which I introduced earlier, are commonly used when talking about things on the atomic scale, for the same reason that we talk about distances between cities in miles or kilometres rather than in inches or centimetres: they fit more comfortably the scale of the things which are being measured.

Let us look at fig. IV-2 again. Every point on the sphere is at exactly the same distance r from the proton. In the ground state of the atom, the wave function changes only when r changes, and therefore it must have the same value at every point on the sphere. Now look at the curve, labelled C, which represents a circle girdling the sphere: its equator, so to speak. Suppose I start at some point on it, and go all the way around the equator till I am back to the same point. Since every point on the equator is at the same distance from the proton, the wave function in this state must have the same value at all these points. I have illustrated this in fig. IV-4. You are looking in this picture at the proton with the equator around it. To help you get a sense of depth, I have marked the front and back of the equator. There is an object like a pin at each of several points on the equator. The height of the pinhead at each point represents the amplitude of the wave function, whose square is the probability of finding the electron there. All these heights are the same, and you should imagine that the probability is the same between the pins also. Again to help you with the perspective, I have shown the pinheads in the front as dark and in the back as clear. What we have here is a wave of constant amplitude around the proton. It is certainly not at all like the waves we see on water, or, for that matter, the waves associated with the electron in a one-dimensional box (fig. III-10). It may help to think of it as a wave with an extremely long wavelength which is tending to infinity.

The probability in the ground state depends only on how far the electron is from the proton, and not on which direction. So we conclude that we find the same probability if we go along any line of latitude or of longitude, so long as it is at the same distance r from the proton.

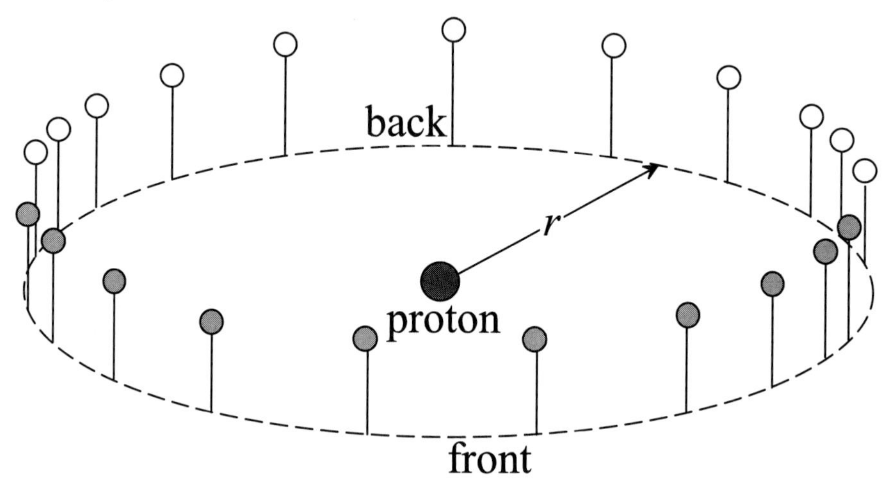

FIGURE IV-4. A depiction of the wave function of the ground state for the electron in the hydrogen atom. Imagine a circle of some radius r at the centre of which sits the proton. The wave function has the same amplitude at every point on this circle, and this value is suggested by the height of the pins stuck around on the circle. The pinheads in the front are shaded dark, to help in the visualisation of depth in the picture: the usual problem of showing three dimensions on a sheet of paper.

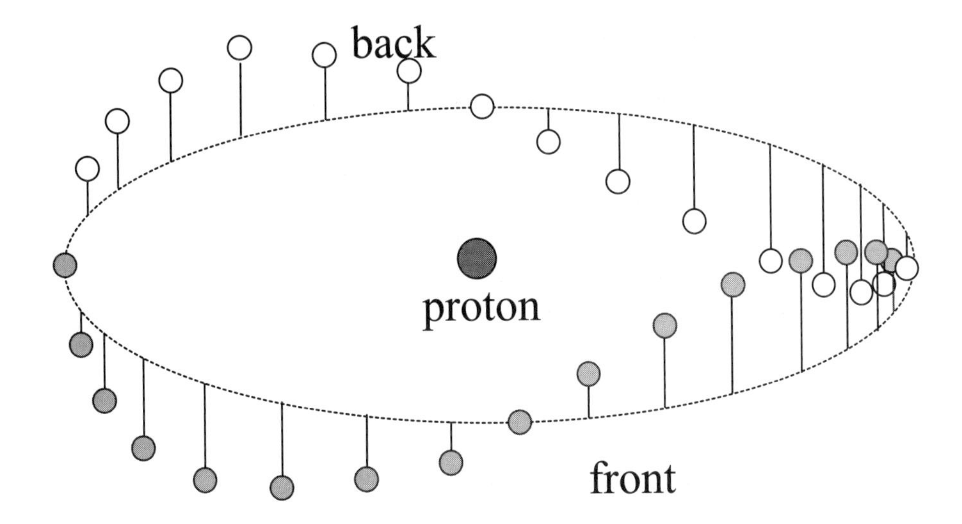

FIGURE IV-5. The wave function for an excited state in the hydrogen atom, going round a circle of longitude with the proton at the centre. Its amplitude is indicated by the height of the pinhead above the circle. The pinheads in the front are shaded dark. This looks more like what one expects from a wave: it has two oscillations in going round the circle.

5 The hydrogen atom: excited states

We now come to the *excited states*, which are the states with energies higher than the ground state energy. We find here that the probability often depends not only on the distance of the electron from the proton, but also on its direction. When I work out the Schroedinger equation, I find that the resulting wave functions (and consequently the probabilities) cannot take any arbitrary form, but can only be one of a discrete set, each with its value for the energy: the energy is quantized. These wave functions, unlike the one for the ground state, have a recognizable wavelike character.

I show in fig. IV-5 how the wave function looks for one of these states, if I follow it along a line of longitude. Since the wave must join smoothly on to itself after going full-circle, its wavelength cannot be any arbitrary value.

QUESTION Why must the wave join smoothly on to itself after going full-circle?

ANSWER Because the probability of finding the electron near any point in space has a unique value in a given quantum state, and cannot change abruptly from point to neighbouring point.

The example I have chosen has exactly two full wavelengths in going around the circle. A similar condition holds if I go around any other line of longitude for this wave function. You can see an analogy with the case of an electron in a one-dimensional box, where also we found that the waves could have only particular wavelengths because they had to fit exactly inside the box. It is this feature that is at the root of the quantization of energy. I can have a wave function with, say, one full wave going around the line of longitude, or two full waves, but nothing in between. Each of these two states has its specified energy, which is one of the discrete energy levels allowed for the atom. This is what we mean by the quantization of energy.

6 Energy levels and quantum numbers

We thus find that the hydrogen atom can be described by any one of a number of wave functions, each of which corresponds to a particular energy of the atom. Some of these energy levels are shown pictorially in fig. IV-6. Let us look first at the right-hand half of this picture. At the bottom is a horizontal line labelled –13.6 eV, representing the energy level of the ground state. The energy increases (i.e. becomes less negative) as we go up from this line, and the next line, at –3.4 eV, stands for the energy level of the first excited state. Yet higher up is the energy level of the second excited state, at –1.5 eV. The energy difference between adjacent levels continues to become smaller and smaller as we go to higher excited states, which I have not shown in the picture. Finally we reach the level labelled 0, whose energy is zero and which represents the electron far away and free from the proton. This is the *ionised state* of the atom. Any further increase in energy would produce motion of the electron and proton relative to each other.

QUESTION How much energy must I feed into the atom in the ground state to put it in its first excited state?

ANSWER Exactly enough to raise its energy from –13.6 eV to –3.4 eV, that is to say 13.6 minus 3.4 or 10.2 eV.

QUESTION What happens if the atom changes from the second excited state (–1.5 eV) to the first excited state (–3.4 eV)?

ANSWER The atom loses an amount of energy equal to 3.4 minus 1.5 or 1.9 eV. By the conservation of energy, this energy has to appear somewhere else: it could appear for example as a photon of energy 1.9 eV.

QUESTION Can I get the atom in the ground state to absorb a photon of 10.0 eV?

ANSWER No. If the atom were to absorb an energy of 10.0 eV, then its final state should have an energy of –3.6 eV. No state with this energy exists for the atom, which therefore cannot absorb it.

Let us now turn to the left half of fig. IV-6. We see three columns of numbers, labelled at the top n, l, and E respectively. The numbers under E (for energy) are the

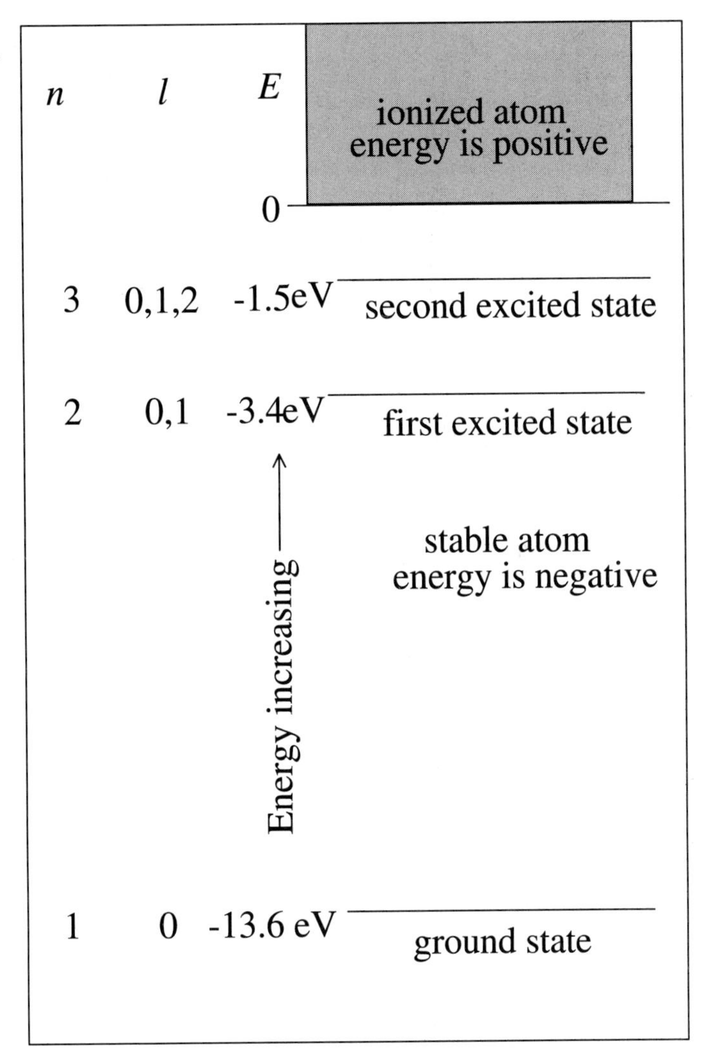

FIGURE IV-6. On the right are shown the quantized discrete energy levels of the hydrogen atom as well as the continuum of energy levels when it is ionized. On the left are three columns of numbers assigned to the discrete energy levels. The columns labelled n and l show the quantum numbers (explained in the text), and the column labelled E the energy, for each level.

energies in eV of the levels, as we have already seen. The numbers in the columns under n and l are called quantum numbers, and are an essential feature of the quantum nature of the atom. You are to read these numbers horizontally. I mean by this that, for example, the second excited state has an energy of −1.5 eV, its n-quantum number is 3, and its l-quantum number can have the values 0, 1, or 2.

We already met a quantum number when we looked at an electron in a one-dimensional box in chapter III. The number n that appeared in the formula for the energy on page 58 is a quantum number. When I work out the wave functions for a given problem, say the hydrogen atom, I find that quantum numbers appear in these functions. These numbers are allowed to have only certain values that are determined by the conditions of the problem, and they set the allowed values of the energy and other such quantities. In the example of the electron in the one-dimensional box, the quantum number n was allowed to take only integer values 1, 2, 3, ... because the electron wave had to fit exactly inside the box, and this number in turn fixed the energy of the electron. I shall now say something more about quantum numbers, using the ground and first excited states as illustrations.

We see from the figure that the ground state has quantum numbers n equal to 1 and l equal to zero. Its wave function contains these two numbers, and I can use a convenient shorthand to write it as $\psi(n = 1, l = 0)$. This is not the end of the story. You remember that the electron has not only a charge, it also acts like a tiny magnet that can point in two opposite directions in the presence of a magnet, and each of these orientations is a different state of the electron with its own wave function. So let me introduce a new quantum number with the symbol s for this property of the electron, and give it the value $+\frac{1}{2}$ for pointing one way and $-\frac{1}{2}$ for pointing the other way. The reason for choosing these peculiar numbers is that it then simplifies other formulas which the physicist uses. Then we can write down the corresponding wave functions as

$$\psi(n = 1, l = 0, s = +\tfrac{1}{2})$$

and

$$\psi(n = 1, l = 0, s = -\tfrac{1}{2}).$$

The energy is the same, −13.6 eV, for these two wavefunctions, and is set by the value of n alone. Now if you are getting a bit alarmed by the complicated symbol for the wave function I have just used, I ask you to think of it just as a shorthand way of reminding yourself that the wave function for the electron has the values of the quantum numbers shown inside the parentheses.

Thus we find that there are two different wave functions for the ground state with exactly the same energy. You may think this a little peculiar; it would be quite a coincidence to have two wave functions that look different and yet have exactly the same energy. Consider this example: a square that is four centimetres on the side looks quite different from a rectangle that is sixteen centimetres by one centimetre, but they have exactly the same area. So it is not difficult to imagine that there can be two, or even more, things that are different from one another in certain ways and yet have one property that is exactly the same for all of them. Physicists have a name for this occurrence of two or more wave functions having the same energy. They say that the energy level is *degenerate.* I have a friend, not a physicist, who giggles each time I use the term in a physics context. She says it conjures up for her an image of a person with bloodshot eyes and ragged hair: a degenerate. This is one more example of the special meaning in physics that we attach to many words of common usage.

You recall that for the electron in a box we had a picture of what the wave function looked like: it looked like a wave. Things get more complicated with atoms and even more so when we get to crystals, so that it will not be always possible to picture their wave functions. So we shall sometimes specify a state of the system by saying what values the quantum numbers have in that state, and not worry about what the wave function looks like. You will now see that this is what I did above when I wrote down the wave functions for the ground state of the atom.

I have so far introduced three quantum numbers:

$$n = 1, 2, 3, \dots,$$
$$l = 0, 1, 2, \dots,$$
$$\text{and } s = +\tfrac{1}{2}, -\tfrac{1}{2}.$$

We have seen that the value of n in the wave function tells me the energy level to which it belongs: 1 for the ground state, 2 for the first excited state, and so on. The two possible values of the quantum number s distinguish between the two opposite orientations of the electron-magnet. Now what about the quantum number l? It contains information about another property of the atom, namely its angular momentum. This property, you remember from chapter III, is connected with rotational motion. Furthermore, the quantum number l contains hidden in it another quantum number that I shall call m. I explain all this by looking now at the first excited state.

This first excited state, as I see from fig. IV-6, has $n = 2$, while l can be either 0 or 1 and s can be either $-\frac{1}{2}$ or $+\frac{1}{2}$. The possible values of the quantum number m are set by what value l has. When l is zero, m is also zero. When l has the value 1, m can take any of the three values +1, 0, or −1. I show in table IV-1 the different combinations of possible quantum numbers.

Table IV-1. *Quantum numbers for the first excited state*

n	l	m	s
2	0	0	$+\frac{1}{2}$
2	0	0	$-\frac{1}{2}$
2	1	+1	$+\frac{1}{2}$
2	1	+1	$-\frac{1}{2}$
2	1	0	$+\frac{1}{2}$
2	1	0	$-\frac{1}{2}$
2	1	−1	$+\frac{1}{2}$
2	1	−1	$-\frac{1}{2}$

Each row in the table is a set of possible quantum numbers for the first excited state. Each set has its wave function, and no two sets are identical. We see that there are eight such wave functions, each of them having exactly the same energy of −3.4 eV. This state is even more degenerate than the ground state, which had only two wave functions.

You may have begun to see a pattern emerging about how the quantum numbers n, l and m are interrelated. It is convenient to express this pattern using algebraic symbols, and this I do now. For a given value of the quantum number n, l can take any of the integer values between 0 and n–1:

$$l = 0, 1, 2, \dots, n-1.$$

With each value of l, m takes any of the integer values from $+l$ to $-l$:

$$m = +l, +l-1, \ldots, -l+1, -l.$$

Finally, with each combination of values of l and m, s takes on the values $-\frac{1}{2}$ and $+\frac{1}{2}$.

QUESTION How many different wave functions are there for the excited state with $n = 3$? And for $n = 4$?

ANSWER When n is equal to 3, l can take the values 0, 1, 2. Then I can write down the possible values of l and m as follows:

$l = 0$ and $m = 0$;

$l = 1$ and $m = +1, 0$, or -1;

$l = 2$ and $m = +2, +1, 0, -1$, or -2.

I thus have a total of 1 + 3 + 5, or nine possible combinations of l and m. Each of these can have s equal to either $+\frac{1}{2}$ or $-\frac{1}{2}$, so that I end up with a grand total of two times nine or eighteen different wave functions for the state with $n = 3$.

When $n = 4$, l can also take the value 3, besides the above values. You can use the same method as before to work out that there are 14 wave functions with $l = 3$, so that there will be 14 + 18 or 32 different wave functions for $n = 4$.

I said that the quantum number l is related to the atom's angular momentum, and we have seen that m is very closely connected with l. If I did an experiment to measure the angular momentum of the atom along some particular direction, I would find that it can only have one of the values given by m multiplied by h and divided by twice π, or more compactly expressed,

$$\frac{mh}{2\pi}$$

where π is a number approximately equal to 3.14. You may remember from your geometry in school that π is the number you get by dividing the circumference of a circle by its diameter, and is exactly the same for all circles, whatever their size. With this result, we now meet a new and extremely important consequence of quantum mechanics, namely that the angular momentum can only have certain values related to the Planck constant, and cannot take on any values that lie in between. The angular momentum is quantized.

The electron not only has an electric charge $-e$ and behaves like a tiny magnet, but also has an angular momentum as though it were spinning like a top. When I measure this angular momentum, I find that it is also quantized: it can only have one of the two values

$$\frac{sh}{2\pi}$$

where s is either $+\frac{1}{2}$ or $-\frac{1}{2}$. In words: the measured angular momentum of the electron, which is a vector, can point in either of two opposite directions with the same magnitude. You can see now why we chose the particular values of plus or minus one-half for the quantum number s.

7 The electronic structure of atoms

I summarize now the picture we have developed so far. The hydrogen atom has a set of energy levels. Each level can have more than one wave function: it is degenerate. We can label each wave function by the values of its quantum numbers n, l, m, and s. Now the question is, in which of these states would a hydrogen atom find itself?

The answer is that the atom will be in its ground state. This is so because, if it begins in an excited state, any slight disturbance will cause it to emit a photon (or in some other way lose energy to the surroundings) and change to the ground state. The energy so lost by the atom will be exactly equal to the difference in energies of the two states, because the total energy must be conserved. The ground state is the stable state of the atom, because it has no state of lower energy to which it can go.

The next simplest atom must have two electrons with total charge $-2e$, and therefore a nucleus with charge $+2e$, because the atom as a whole is neutral with zero total charge. Its energy level picture will be rather different from the one in fig. IV-6, because now we have to think of the quantum numbers for two electrons. The energies of the levels will be different from what they were in the hydrogen atom, because the charge on the nucleus is $+2e$ instead of $+e$. What we have now is an atom of helium, which you know as the gas used to fill balloons on fairgrounds and for children's birthday parties. The two electrons go into the ground state, and their quantum numbers are as shown below:

$$\begin{array}{ll} \text{First electron:} & n=1,\ l=0,\ s=-\tfrac{1}{2} \\ \text{Second electron:} & n=1,\ l=0,\ s=+\tfrac{1}{2} \end{array}$$

You will note that not all the quantum numbers for the two electrons are the same. One other point: the nucleus of the helium atom consists of two protons each with a charge $+e$ and two neutrons with no charge.

The description of the helium atom which I have given raises two questions whose answers introduce us to some very important new ideas. The first question is, since I can balance the charge of $-2e$ of the two electrons with just two protons each with a charge $+e$, why do I want to throw in two neutrons also which, after all, have no charge? Well, you remember that two like charges repel each other, so that if I had just two protons in the nucleus, they would fly apart. It turns out that there is another kind of force that acts only among neutrons and protons, and this force attracts them to one another very strongly. We call this the *nuclear force*. This attractive nuclear force, however, is still not strong enough to overcome the repulsive electric force that is trying to push the protons apart. The two neutrons increase the attractive nuclear force further, without affecting the electric force (because they are electrically neutral), enough to hold the nucleus together. A study of the nuclear force is at the heart of nuclear physics, which is one of the branches of physics. We shall not dwell further on it, since what goes on inside the nucleus is not relevant to those properties of matter that we want to understand in this book.

And now for the second question: When I put the two electrons in the ground state, why did I not put them both in a state with the same value for the s quantum number instead of putting one of them in the $s = -\frac{1}{2}$ state and the other in the $s = +\frac{1}{2}$ state? The answer lies in a peculiar property of electrons that I have not mentioned so far. This is that no two electrons can have all of their quantum numbers exactly the same. A given state, with certain values for its quantum numbers, can be occupied by one, and only one, electron. It is as if an electron, sitting in a given state, excludes any other electron from getting into the same state. This so-called *exclusion principle* was discovered by the physicist Wolfgang Pauli, and is named for him. I shall say more about it in chapter V. We shall use it now to form a picture of the wave functions for electrons in a given atom.

To be specific, let us consider an atom of carbon. Its nucleus consists of six protons and six neutrons. To keep it electrically neutral, it must have six electrons surrounding the nucleus. We have seen that the $n = 1$ level has two states, and, from table IV-1, the $n = 2$ level has two states with $l = 0$ and six states with $l = 1$, with no two states having all quantum numbers the same. So we start by putting two electrons in the $n = 1$ level, and this fills it up. We then put two electrons in the $n = 2$, $l = 0$ level, filling it up. The last two electrons go into the $n = 2$, $l = 1$ level and the result is an atom of carbon. You can see that the exclusion principle has been satisfied by this distribution of electrons.

I hope you are beginning to see now the picture of an atom given by quantum mechanics. Let me describe it, putting together the points made in this chapter. All the atoms of each chemical element are identical. Each atom has a nucleus with a certain number Z of protons and a number N of neutrons. I show the values of Z and N for some elements in table IV-2. Surrounding the nucleus are Z electrons, so that the atom is electrically neutral; the charge $+Ze$ on the nucleus is exactly cancelled by the charge $-Ze$ of the Z electrons. The electrons are to be found in the successive energy levels, starting at the ground level. According to the Pauli exclusion principle, no two electrons will have all the quantum numbers the same. The value of Z, the number of protons in the nucleus (which is the same as the number of electrons in the atom), identifies the element uniquely; each element has a different value of Z.

Table IV-2. *The number of protons, electrons and neutrons in some atoms*

Element	Z (protons, electrons)	N (neutrons)
hydrogen	1	0
helium	2	2
carbon	6	6
uranium	92	146

If you look at fig. IV-3 again, you will be reminded that there is very little probability of finding the electron at a distance of more than about an angstrom from the proton in the hydrogen atom. We can say that the size of that atom is about one angstrom. The electrons in the other atoms go into successively higher excited states. In these states, the electron has a somewhat higher probability of being further away from the nucleus than in the ground state, but even then not by very much. Thus it turns out that in the atoms of all the elements, there is very little probability of finding any electron further from the nucleus than a couple of angstroms. Even though the number of electrons in atoms of different elements varies between one and about one hundred, the sizes of the atoms are all a few angstroms across. The nucleus itself is far, far smaller than this. Its size depends on how many neutrons and protons it has, and in all cases is a few times 10^{-5} angstroms.

I remind you again that no two of these electrons have all of their quantum numbers the same. You must imagine that there exists a wave function that describes the probability of finding any one of these electrons at any specified region around the nucleus, and is such that the Pauli exclusion principle is obeyed. I am afraid I cannot draw a picture of this wave function for, say, a carbon atom in the way I was able to for the much simpler case of a single electron trapped in a box in chapter III. This is where mathematics comes to the aid of the physicist. She will say that she does indeed have a picture of the atom, but it is expressed in mathematical formulas and not in lines and shadings on a sheet of paper. We shall see as we go along that we can make much headway even without all that mathematics.

All the things we see around us are made up of atoms. The only difference between atoms of different elements is in the number of electrons around the nucleus and the numbers of protons and neutrons in the nucleus. This might appear to be too trivial a difference to account for the amazing variety of properties that things around us have. A lump of coal, which is nearly pure carbon, is not likely to be confused for a lump of silver, or for a diamond which is also pure carbon. My main task in the rest of this book is to explain how these properties follow from the way in which the electronic structure of the atoms is modified when I put them together to form a solid.

The picture of an atom that we have arrived at is not at all what one might have thought, namely a tiny hard ball a few angstroms across in size. We can make this dramatically evident if we imagine that we have somehow magnified an atom so that the nucleus measures a centimetre across. Then the electrons, numbering at most about a hundred, will be distributed in a sphere of diameter about one kilometre. The cloud of electrons around the nucleus is a very thin cloud indeed; most of the atom looks like empty space. It would appear that one atom should be able to pass right through a second atom and keep going. This is not what happens in reality. When I put atoms together to form a solid, they behave as if they were hard impenetrable balls stacked together. To take a more immediate example, I am sitting comfortably on a chair as I write this; my atoms, and therefore I, do not just simply pass through the atoms in the chair. Why is this so?

The answer lies in the Pauli exclusion principle, which says that one, and only one, electron can occupy a given state with a given set of quantum numbers. A second electron must necessarily go to another state with a different set of quantum numbers. Now let us see what happens when I try to bring two atoms of carbon, say, together. Figure IV-7 shows the two atoms some distance apart. You can think of it as a kind of snapshot. You see in each atom the nucleus at the centre and six electrons in the surrounding grey space, which represents the sphere within which each of the electrons is likely to be found. Please note that this and the following picture are not drawn to scale. Each electron is in one of the six possible states of lowest energy. Now suppose I try, as in fig. IV-8, to bring the two atoms together so that the two grey spheres overlap. Then in the overlapping region I am forced to have pairs of electrons, one originally from each atom, trying to occupy the same state. This, we know, is prohibited by the Pauli exclusion principle, which allows only one electron to be in each state. The consequence is that the atoms cannot penetrate into each other, and my chair supports me and does not let me pass right through it.

I have said that the number of electrons in an atom uniquely specifies which element it is: one for hydrogen, six for carbon, and so on. This is also the number of protons in the nucleus. There are also some neutrons in the nucleus, which have no electric charge. So their number is not fixed, but can vary a bit without affecting the electrical

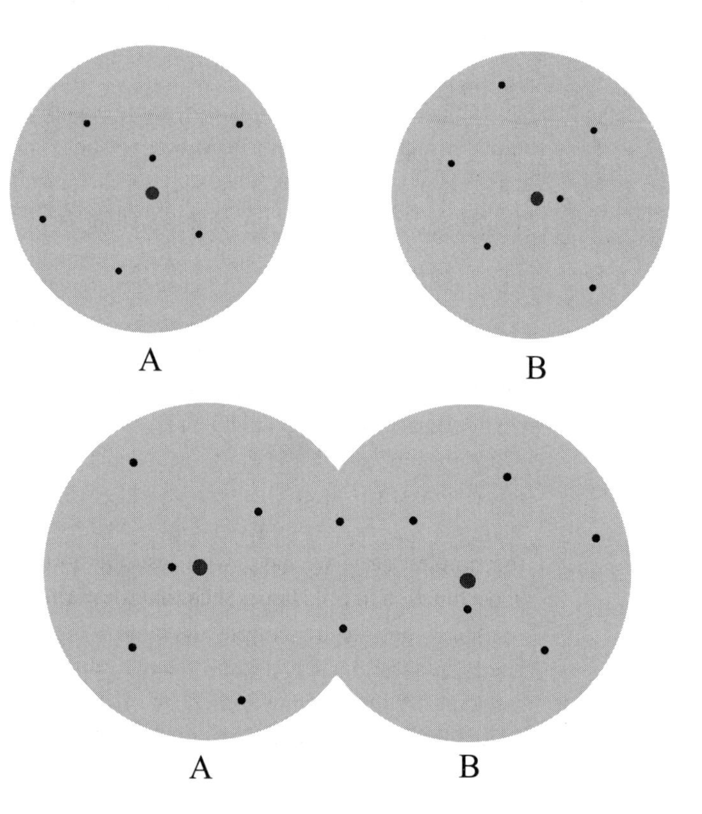

FIGURE IV-7. Two carbon atoms A and B at some distance apart. The nucleus is at the centre of each atom, and six electrons surround it. The grey shading around the nucleus indicates the space within which the six electrons are almost certain to be, according to their wave functions.

FIGURE IV-8. The two carbon atoms of fig. IV-7 now brought close enough to each other that their wave functions begin to overlap, thus requiring that two electrons, one from each atom, have the same wave function. This is not possible, because of the Pauli principle, and so the atoms cannot penetrate each other.

neutrality of the atom or changing it to a different element. I shall take the simplest example, an atom with one electron which therefore must have a nucleus with charge $+e$. I accomplish this by having a nucleus of just one proton, and this of course is the hydrogen atom. There also exist two other forms of hydrogen, rare to be sure, with the nucleus of one consisting of one proton and one neutron, and the other of one proton and two neutrons. They are called deuterium, or sometimes heavy hydrogen, and tritium respectively. Many of you have heard of isotopes. Ordinary hydrogen, deuterium and tritium constitute three isotopes of the same element. There are isotopic forms of all the elements. Isotopes are of interest when one studies the physics of the atomic nucleus, and they have found applications in certain fields of medicine, technology, and weaponry. For our purposes in this book, however, we can ignore their existence, for none of the characteristics of a solid that we want to understand depends noticeably on its isotopic composition. Well, almost none; in superconductivity, the isotopic nature of the material plays a role that gives an important clue to our understanding of the phenomenon, as we shall see in chapter XII.

8 Summary

The gravitational attraction between the sun and the planets is what holds the solar system together. While this attraction certainly exists between the electrons and the nucleus in an atom too, it is far too weak to hold the atom together. It is a different force, the electric force, which is at work in the atom. Section 2 introduces you to this force, and relates it to a property of electrons and protons called electric charge. This charge is the source of the electric force, just as mass is the source of the gravitational force. Unlike gravitation which always pulls two bodies towards each other, the electric force either pushes them apart (two electrons or two protons) or pulls them together (an electron and a proton). Because of this, we say that a charge can be negative or positive, and like charges repel and unlike charges attract each other.

In section 3 we see an important consequence of the wave aspect of the electron: it cannot fall into and remain trapped inside the nucleus. If it did, it would have such an enormous kinetic energy of motion that it would be shot out of the nucleus right away.

The electron in a one-dimensional box has a state of lowest energy, with a wave having just one trough (or one crest) fitting exactly inside the box. We look in section 4 at the analogous lowest energy state, the ground state, of the hydrogen atom. This is the simplest atom we have, with just one proton and one electron. We find that the wave function in this state is such that the electron is most likely to be found very close to the proton, and less and less likely as we go further away. The likelihood becomes practically zero at a distance of about an angstrom, or 10^{-8} centimetres, so that we can think of the atom as measuring about an angstrom across.

In section 5 we look at the states of the electron with higher energies, the so-called excited states. Each of these states has a definite energy, and these energies are separated by gaps. The electron is allowed to have only these energies, and nothing in between. This is what we call the quantization of energy.

We look in section 6 more closely at the energy levels of the atom. Each of these levels is degenerate, i.e. there is more than one wave function that corresponds to this energy level. This means that the electron can have different probability distributions around the nucleus, all of them having the same energy. We introduce quantum numbers *n*. *l*, *m*, and *s* for the wave functions. Each wave function is labelled by the particular numerical values which these quantum numbers assume in it. The number *n* is related to the energy of the electron in that state, *l* and *m* to that part of the angular momentum of the atom which arises from the way the electron wave is distributed around the proton, and *s* to the angular momentum of the electron by itself. We find that not only the energy but also the angular momentum is quantized. The allowed values of the angular momentum are determined by the Planck constant *h*, and the numerical values of *m* and *s*. The atom is usually found in its lowest energy state, namely the ground state. It can be put into an excited state by giving it an amount of energy that is exactly equal to the difference in energies between the two states, for example by making it absorb a photon of this energy.

Using the scheme of energy levels and quantum numbers, we see in section 7 how the atoms of other elements like helium, carbon and uranium are built up. The nucleus now has several protons and neutrons, and the number of protons determines uniquely which element it is. Surrounding the nucleus is a number of electrons equal to the number of protons. Each electron has its wave function with a set of quantum numbers which are different from those of any other electron. No two electrons can have exactly the same set of values for the quantum numbers. This is an example of the Pauli exclusion principle. One can say that the electrons do not seem to be very friendly characters, each preventing any other electron from entering its house.

What I have described so far is an atom in isolation from the rest of the world, not disturbed in any way by all the surrounding atoms because they are too far away. The

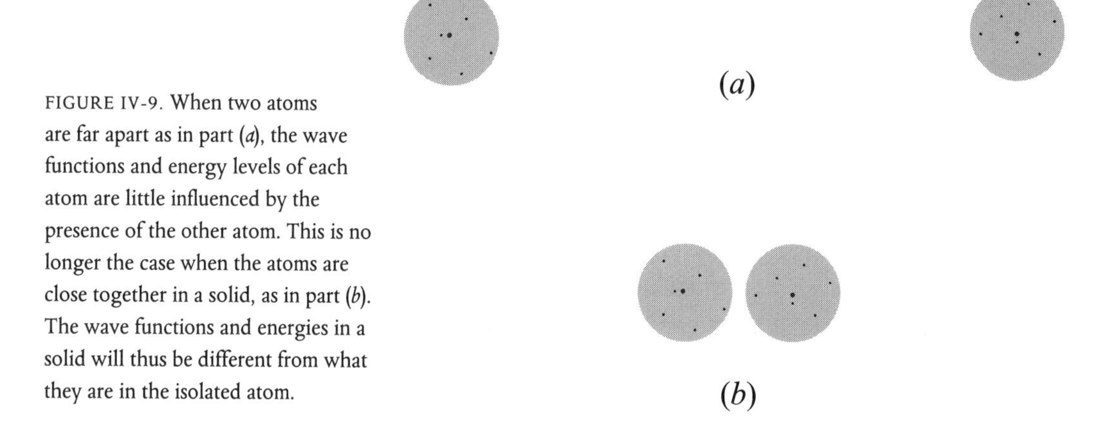

FIGURE IV-9. When two atoms are far apart as in part (*a*), the wave functions and energy levels of each atom are little influenced by the presence of the other atom. This is no longer the case when the atoms are close together in a solid, as in part (*b*). The wave functions and energies in a solid will thus be different from what they are in the isolated atom.

atoms in a gas are a very close approximation to this state of affairs. Their average distance apart is much larger than their size. We remember that it is the electrical force that is the important one in atoms. Now look at the two atoms in fig. IV-9, part (*a*). I have shown them as being very far apart in relation to their size. Then the electrons and the nucleus of each atom are all at about the same distance from the other atom. Each atom feels no electrical force due to the other atom, because the force due to the electrons (negative charge) will be cancelled by that due to the nucleus (equal positive charge). So each atom is unaware of the other's existence, so to speak. Now look at the two atoms in part (*b*) of fig. IV-9. Their distance apart is comparable to their size. You can see qualitatively that the electrons and the nucleus of each atom must each feel the influence of the other atom, and we can no longer think of isolated atoms. The atoms in solids are at distances apart which are like that of the atoms in part (*b*): the lattice spacing in crystals is a few angstroms, and is therefore comparable to the size of atoms. We can expect that the wave functions and energy levels in solids will be changed from what they are for the isolated atom. These are the changes we shall look at in chapter VI. We shall then be able to see why solids are the way they are.

Before getting there, there is one other matter we have to deal with. We have seen so far what the quantum-mechanical picture of single atoms looks like. In an actual solid, there are very large numbers of atoms packed rather closely together. Very large numbers indeed: a cube of solid measuring one centimetre along the edge will have something like 10^{21} atoms in it. That means that there will also be comparably large numbers of electrons in the solid. It is obviously a hopeless task to try to find out what each electron is doing, because even the lifetime of the universe would not be enough for the task. We need to invent a new way of dealing with such large numbers of objects. This way is called *statistical physics*, and it is the subject of the next chapter.

V STATISTICAL PHYSICS

1 Introduction

Any object that we see around us, even though it may be very small like a grain of sand or a pin, consists of a very large number of atoms, about 10^{21}. When the atoms are relatively far apart, as in a gas, the electronic structure, meaning the wave functions, of each atom is practically unaffected by the presence of the other atoms. When the atoms come close together in a solid, the wave functions become different from what they were for the isolated atoms. In crystals of metallic elements like silver and copper this change is such that the electrons in the highest energy state of the isolated atom are no longer bound to stay near the corresponding nucleus, but are free to wander around in the whole crystal. This freeing of the outer electrons happens with each atom in the crystal. So we end up with a crystal in which a large number of electrons, 10^{21} or more, are wandering about. These electrons are responsible for many of the metallic properties of copper or silver: the shiny appearance and the ability to carry electric current and to conduct heat easily, for example. We shall see later how this comes about. For the present we need to find a way to describe the properties and behaviour of such large numbers of identical entities like electrons or atoms. We do this with statistics.

The Concise Oxford Dictionary defines the word *statistics* as numerical facts systematically collected. It is in this sense that the word is used when we talk, for example, about weather statistics: the temperatures, rainfall and so on for various cities that are reported in the newspapers every day. A typical report might cover some fifty cities. What if we want to present the same information for a thousand places scattered all over the world? Or a million places? We would then be faced with a hopeless task – quite apart from the question of what use the resulting compilation would be.

Fortunately, we do not have to undertake a similarly daunting task when we consider the behaviour of the 10^{21} or so particles, like electrons or atoms, that are in a typical piece of solid matter or volume of gas or liquid. It turns out that for our purpose of understanding the behaviour of matter we do not need to know what each particle in this huge assembly is doing, but can be satisfied with much less information about groups of them. To explain what this means, I shall first describe a simple example. Figure V-1 shows some information about the age of the populations of two countries: the U.S.A. in the year 1989 and India in 1990. Along the horizontal line at the bottom you see the labels 0 to 9, 10 to 19, and so on, which stand for ages up to 9 years, between 10 and 19 years, … . Above each of these labels are two bars, one for the U.S.A. and the other for India. The length of each bar represents the percentage of the population that is in the corresponding age group. The vertical line on the left of the

FIGURE V-1. The population distribution by age in the U.S.A. and in India in the years 1989 and 1990 respectively. For example, about 14 per cent of people in the U.S.A. and 22 per cent of those in India were between 10 and 19 years old.

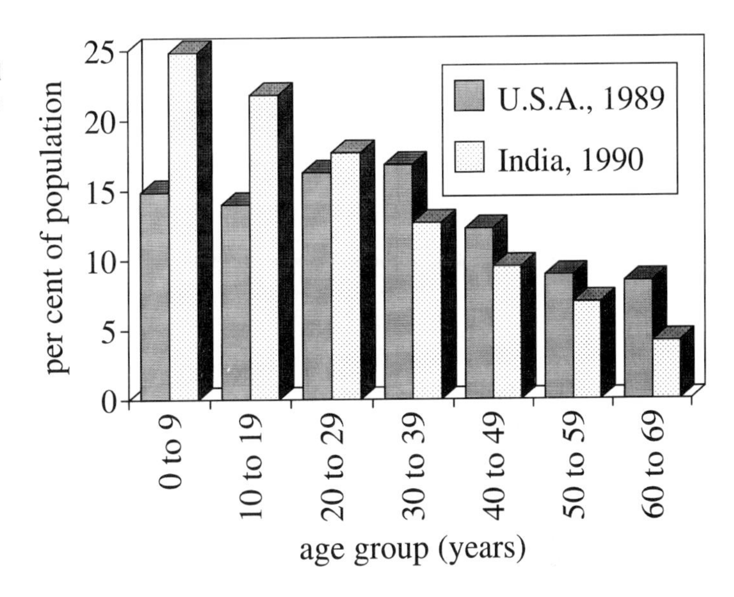

diagram is a scale that enables you to tell how long each of the rectangular bars is, just as an ordinary scale for measuring lengths tells you how long something is in centimetres or inches. For example, the figure shows that about 14 per cent of the population of the U.S.A. and 22 per cent of Indians are between 10 and 19 years old.

You will note that this figure, although it tells us something about the ages of the people, is quite different from a simple, if very long, list of the age of each person in the two populations. It tells us how the people are distributed in different age groups. For this reason we say that fig. V-1 shows the *distribution* of the populations of the U.S.A. and India by age.

QUESTION What are the percentages for the age group from 0 to 9 years in the U.S.A. and India?

ANSWER It is about 15 per cent in the U.S.A, and 25 per cent in India.

There were 248 million people in the U.S.A. and 827 million in India when these counts were made. You can imagine the impossibility of trying to make a list showing the age of each person. It suffices for many purposes just to know what fractions of the population are up to 9 years old, 10 to 19 years old, and so on. Such information would be needed in making plans for education, health care, social security, and other such matters. The picture of the distribution offers something more: it presents in a visually striking manner something about the populations that would be invisible in a bare table showing the age of each person, and not easy to see even in table V-1, which shows the numbers that were used in drawing the figure. I point out the following features of the way the percentage of people in a given age group varies as the age of the group increases:

Table V-1. *Distribution by age groups of populations of the U.S.A. and India*

age group (years)	0–9	10–19	20–29	30–39	40–49	50–59	60–69
per cent (U.S.A.)	14.9	14.0	16.3	16.8	12.2	8.9	8.5
per cent (India)	24.8	21.8	17.7	12.7	9.5	7.0	4.2

1. For the U.S.A., this percentage slowly increases to a maximum value at the 30 to 39 year age group, and then begins to decrease.
2. For India, this percentage starts at a relatively high level for the youngest age group, and then steadily declines as the age increases. The pattern is strikingly different from that of the U.S.A.

One can make sense of these two features if one thinks about the relative affluence of the two countries and the consequence for infant mortality, health care, and nutrition, and also recalls that for some years in the 1950s, the so-called baby-boom years, the U.S.A. had an unusually high birth rate. I shall not dwell further on this, for this book is about physics and not demographics. I do note three features of this way of presenting the information. The first is that we know something about the distribution of the population among different age groups, but nothing about the age of any given individual. The second is that the pattern of the age distribution implicitly says something about the population: its affluence, state of health, and even history. Thirdly, the information in this age distribution can be used for other purposes such as the planning of health care, social services, and so on. There are analogies to these three features in the way we describe a collection of a large number of particles in physics. This description is called *statistical physics.* Before I get to that, I must explain to you what temperature means in terms of the atomic structure of matter. As you will see soon, temperature plays a crucial role in statistical physics.

2 Temperature

Boiling water is hot, and ice is cold. I specify how hot or cold something is by giving its temperature. Water boils in New York at a temperature of 212 degrees Fahrenheit, and freezes to ice at 32 degrees Fahrenheit. In Rome, on the other hand, water boils at 100 degrees Celsius and freezes at 0 degrees Celsius. This does not mean that Italian water is somehow different from American water, but only that people in the two cities use different units with which to measure temperature. In this book we shall follow the practice of physicists, and indeed of most of the world, and use Celsius degrees. We shall write 'twenty degrees celsius' as '20 °C'. There is yet another way of expressing temperature, by adding the number 273 to the degrees Celsius and calling the result as so many *kelvin.* For example, 20 °C would be the same as 293 kelvin, which we shall write as 293 K. You will see in a while that there is some point to expressing temperatures this way.

We get our idea of something being hot or cold through our sense of touch. I now want to connect this idea with what the atoms are doing in a piece of matter. Suppose I take a container filled with helium gas, say, which is at a temperature of 20 °C. The helium atoms are not just standing still, but are whizzing about in all directions at all possible speeds, bumping into one another and with the walls of the container. When two atoms collide, the speed of each will in general be different from what it was before. You can see that it is hopeless to expect to follow the motion of each atom as time goes on. Fortunately, we do not need to do this. Numerous experiments have shown that the number of atoms whose speeds lie in any given range will remain unchanged, so long as the temperature remains unchanged. The atoms whose speeds change to outside this range because of collisions are replaced by an equal number of atoms whose speeds move into this range because of collisions. Now suppose that I heat the container so that it and the gas inside it are at a temperature of 300 °C, say. The atoms will still be moving about at all speeds, but their distribution in different ranges of speeds will be changed by this change of temperature.

Figure V-2 shows the distribution of speeds of the helium atoms at the two temperatures. The picture is based on experiments in which the speeds of groups of atoms were measured, and on a theory worked out by the physicists Ludwig Boltzmann and James Clerk Maxwell. Their predictions agree with the experimental results.

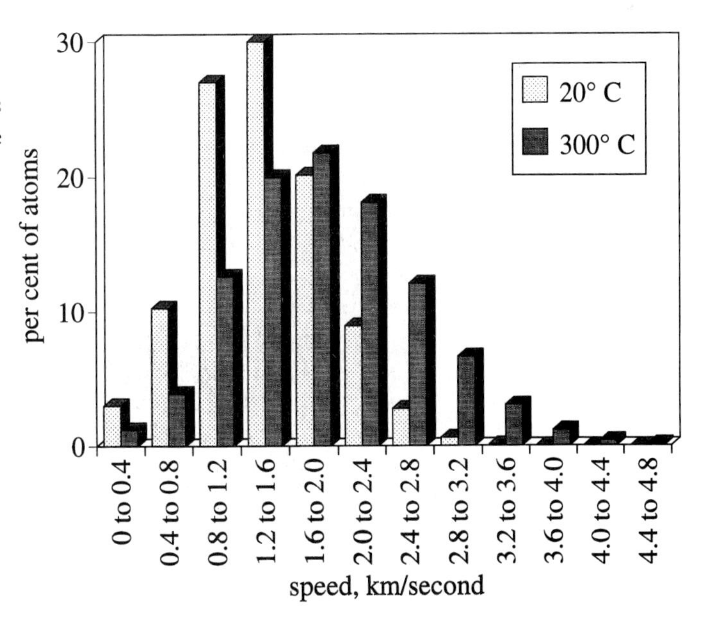

FIGURE V-2. Distribution of atomic speeds in helium gas at 20 °C and 300 °C. The length of each bar shows the per cent of atoms whose speeds lie within the corresponding range shown along the horizontal axis. Each range includes the lower value and ends just below the upper value.

The horizontal axis carries labels showing speeds within the ranges 0 to 0.4, 0.4 to 0.8, ... kilometres per second. I should mention that a speed of 1 km/sec, which is the same as 2250 miles per hour, is somewhat more than what we normally meet; the atoms are indeed speeding about. Above each of these numbers are two bars, one for each of the two temperatures. The length of a bar, measured on the vertical scale at the left, gives the percentage of atoms that have speeds in the corresponding range and at that temperature. For example, I see from the figure that about 30 per cent of the atoms in the gas at 20 °C have speeds between 1.2 and 1.6 km/s, and this drops to 20 per cent when the temperature is raised to 300 °C.

QUESTION What percentage of atoms have speeds in the range from 0.4 to 0.8 km/s?

ANSWER About 10 per cent at 20 °C and 4 per cent at 300 °C.

QUESTION What is the average speed of an atom in the gas at 20 °C?

ANSWER The question is similar to 'If four apples cost one dollar each, and six other apples cost two dollars each, what is the average cost of an apple?' To answer this, we work out the total cost of the ten apples as four times one, plus six times two, or 16 dollars, so that the average cost per apple is 16 divided by 10 or 1.60 dollars. We can work out the average speed in the same way, substituting atoms for apples and speed for cost. Suppose we have 100 atoms in the gas. The actual number will not matter, because we are working only with percentages. We find from the figure that about 27 per cent, or 27 of them, have speeds between 0.8 and 1.2 km/s. We say approximately that each has a speed of 1.0 km/s, and we multiply 27 by 1.0 to give us the number 27.0. We do the same thing with all the speed ranges, add up the resulting numbers, and divide by 100, which is the total number of atoms, to get the average speed per atom. When you work it out, you will find that the average speed is 1.2 km/s. We could have expected the answer to be about this, because we see from the figure that a large fraction of the atoms have speeds around 1.2 km/s. If more than half the apples I bought cost a dollar each, and there was not too much spread in the prices of the rest, then I would expect the average cost per apple to be about a dollar.

I repeat that even though the speed of a given atom may be changing thousands of times per second, the distribution of speeds is unchanged so long as the temperature of the gas remains constant. One might wonder whether there is a deeper connection between the two quantities. There is a slight problem: temperature is just one number, so many kelvin or degrees Celsius, whereas the distribution of speeds is many numbers. So we need to distil some single number out of the distribution that can be meaningfully related to the temperature.

We start the search for this number by looking at some features of fig. V-2. While there are atoms moving at all speeds from practically zero to more than 4.4 km/s, the percentage reaches a maximum value for speeds in the range between 1.2 and 1.6 km/s, and drops off at lower as well as higher speeds. There are some atoms moving at speeds much greater than 4.4 km/s, but they are such a small percentage of the total number of atoms that they do not show up on the scale of this picture.

Now look at the effect of heating up the gas on the distribution of atomic speeds. You can see that the distribution, although it has the same qualitative shape, is quantitatively different. Raising the temperature leaves a smaller fraction of atoms at the lower speeds and increases the fraction of atoms at higher speeds. An appreciable percentage of atoms is now moving at high speeds at which there were practically no atoms at the lower temperature. The range of speeds at which the maximum percentage of atoms occurs is now at 1.6 to 2.0 km/s, while it was at 1.2 to 1.6 km/s at 20 °C. We can calculate the average speed of an atom for the distribution at 300 °C, and find that it is 1.7 km/s, distinctly higher than the value of 1.2 km/s at 20 °C. It would appear that the average speed is a possible candidate for the single number we are looking for to relate to the temperature. It turns out that there is an even better candidate, namely the average energy of an atom in the gas.

3 Temperature and energy

When the container of helium gas was heated from 20 °C to 300 °C, we see from fig. V-2 that the only changes were that there were fewer slower atoms and more faster atoms, and the range of speeds in which the highest percentage of atoms found themselves was higher. As a consequence, the average speed increased from 1.2 km/s at 293 K to 1.7 km/s at 573 K. Notice that I have now changed to kelvin from degrees Celsius, and for good reason too, as you will soon see. From the foregoing observations it seems reasonable to conclude that the temperature must somehow be connected with the state of motion of the atoms, which was the only thing that changed when the gas was heated up. So let us try to find out what this connection might be. To do this, we need to recall some things that we learnt about the motion of particles in chapter III, section 3 (page 37).

We learnt there that a moving atom with mass M and velocity $\mathbf{v}$ has a momentum $\mathbf{p}$ and kinetic energy K which are given by the formulas

$$\mathbf{p} = M\mathbf{v}$$

and

$$K = \frac{Mv^2}{2}$$

where $\mathbf{v}$ is the velocity (with magnitude v) of the atom. We get the total momentum and total kinetic energy of the gas by adding up for all the atoms their individual momenta (remember that they are vectors) and kinetic energies (which are scalars, and always positive). The atoms at any given instant of time will be travelling in every possible direction inside the container, so that the momentum of one atom going in a particular direction will be cancelled by the momentum of some other atom going in the opposite direction with the same speed. This results in the total momentum of the gas being zero, and it remains zero, by the same reasoning, when the gas is heated up. The total momentum of the gas is unchanged when the temperature changes, and so there can be no direct connection between the two.

Things are different with the kinetic energy of the gas. We see from the formula for K that increasing the speed increases K even faster: doubling the speed quadruples the kinetic energy. So the raising of the temperature, resulting in a speeding up of the atoms, causes the total energy (which we shall call E) of the gas to increase. It is tempting to say that the temperature is somehow related to the total energy of the gas. But not so fast! If I took two identical containers of the gas at the same temperature and put them together, the resulting object would have twice the energy of the separate containers, but the temperature would be unchanged.

So I cannot relate the total energy of the gas to its temperature. There are other quantities that are unchanged in the above experiment of putting two containers together. One of them is the average speed of an atom. Another is the average kinetic energy of an atom. I have calculated the latter quantity using the same method as for the average speed, and find that the average kinetic energy is

$$M \times 8.5 \times 10^9 \text{ ergs at } 293\text{ K}$$

and

$$M \times 17.0 \times 10^9 \text{ ergs at } 593\text{ K}.$$

The letter M stands for the mass of the helium atom expressed in grams. The average speed and the average energy depend only on the temperature of the container of gas, and not on how many atoms there are.

The average speed was 1.2 km/s at 293 K and 1.7 km/s at 593 K. The speed does increase with an increase in temperature, but not exactly in proportion. A doubling of the temperature causes a much smaller increase of the average speed. However, when the temperature is almost doubled from 293 K to 593 K, the average energy is also doubled. There exists a simple proportionality between temperature and average energy. You can see now why I chose to measure the temperature in kelvin. It is only then that this simple proportionality appears. We have verified this proportionality for helium gas at two temperatures now. Much work with many gases at many temperatures has confirmed this proportionality of the temperature to the average energy of the atoms or molecules in the gas.

We have thus established a direct connection between the temperature and the average kinetic energy of the atoms in a gas. Let us now think a bit about what this means. You would measure the temperature of the gas by sticking a thermometer in it. The temperature is a property of the gas as a whole, is the same everywhere in it, and the act of measuring it would seem to have nothing to do with the fact that the gas is composed of atoms. Now we see that measuring the temperature is nothing but, as it were, measuring the average energy of the atoms in the gas. We have found a connection between a macroscopic property, the temperature, with an atomistic property, the average energy of atoms. The construction of such connections between features of the macroscopic world around us and their microscopic atomic structure, is the task of the field of statistical physics.

4 The Kelvin scale and the absolute zero

We have seen that the average kinetic energy of an atom in a gas is proportional to its temperature. We can write this as a formula:

$$K = \tfrac{3}{2} k_B T$$

where K is the kinetic energy, T is the temperature in kelvin, and k_B is a constant called the *Boltzmann constant,* after the physicist Ludwig Boltzmann. It has the value

$$k_B = 1.38 \times 10^{-16} \text{ ergs per degree.}$$

The factor $\frac{3}{2}$ in the formula for K is there because of the value chosen for the constant k_B. We see that doubling the temperature doubles the kinetic energy. If we go in the other direction and halve the temperature, then the kinetic energy is also halved and the speeds of the atoms are lower. As we keep reducing the temperature, K gets smaller and smaller. Is there going to be an end to this? Yes, because the lowest energy an atom can have is its zero-point energy, which is a consequence of the wave aspect of the atom inside a container, just as the particle in a one-dimensional box (chapter III) has a finite zero-point energy. This must mean, from the formula above, that the energy of motion (called the thermal energy) which is to be attributed to temperature is zero at a temperature of zero K, and the atoms are left with just their zero-point energy. We shall call this the absolute zero of temperature, 0 K. From the way we defined the Kelvin temperature scale (sometimes also called the absolute temperature scale), we see that the absolute zero is at −273 °C. This is the lowest possible temperature that could exist anywhere at all, corresponding to all the atoms being in their lowest energy state.

5 The temperature of crystals

In the preceding section, I talked glibly about cooling the gas until it reaches the absolute zero, where the atoms have no thermal energy. Well, this does not quite agree with reality. If I take a container of some gas and cool it to lower and yet lower temperatures, then at some temperature (called the boiling point) it will condense into a liquid. On further cooling, it reaches a temperature (the freezing point) at which it becomes a solid. An example: I imagine a container at a temperature of 105 °C filled with steam having the same pressure as the air around us. As the container cools down, the steam will condense to water at 100 °C, and on further cooling, the water will freeze to ice at 0 °C. The motions of atoms in the solid and the liquid will be somewhat different from each other, and each will certainly be very different from that in the gas. So what if anything will survive of the simple relation between temperature and kinetic energy that we found in the gas?

To begin with, it seems reasonable that the temperature in the liquid and the solid must still be connected with the energy of motion of the atoms. To see this, let us think of the gas a bit above the boiling point. Its atoms are moving about at various speeds as we saw, some of them quite large. Now if the temperature drops a little, enough to cause the gas to condense into a liquid, the atoms will have lower speeds and will be bouncing off each other more than they did in the gas because the atoms are closer together. The atomic motions will be much more complicated than in the gas, and it is not possible to present a picture of the distribution of speeds in the same way as in fig. V-2. We can imagine that there is still some distribution which is characteristic of the temperature, and that this distribution shifts to lower speeds as the temperature is lowered.

QUESTION Why should not the distribution of speeds in the liquid shift to higher speeds as the temperature is lowered?

ANSWER Well, if this happens each time we lower the temperature, the distribution will soon be back up at those speeds where we should have gas, and not liquid. Of course, the liquid does not become gas when the temperature is lowered; rather, it will freeze into a solid.

Now let us look more closely at a solid, a crystal to be specific, for that is what we are mostly concerned with in this book. You will remember from chapter II that the atoms in a crystal are in an orderly arrangement at the lattice points. Suppose the crystal is at the absolute zero of temperature, so that it has the lowest possible energy. This would be the case if the atoms are just sitting at the lattice points and not moving at all, for then the kinetic energy of each atom would be zero. However, the wave nature of the atoms does not permit this situation to be realized. Each atom in the crystal is effectively trapped inside a small finite volume around its lattice point. The surrounding atoms block it from leaving this volume. So, just as the particle trapped in a one-dimensional box has a ground-state energy that is not zero, the atom in a crystal has a non-zero lowest energy, its zero-point energy.

I should mention here that nature is such that we cannot cool something down to the absolute zero by any method that has been used or even can be imagined. The closer we get to 0 K, the more difficult it becomes to get any closer. The unattainability of the absolute zero is not a failure of our ingenuity, but rather a consequence of the wave nature of matter. It is still all right to think of the crystal at 0 K as an idealization.

Now suppose I raise the temperature of the crystal by adding energy to it. This energy is taken up by the atoms and produces some sort of motion in them. If the crystal is to remain a crystal, each atom cannot wander away from its lattice site. It can only

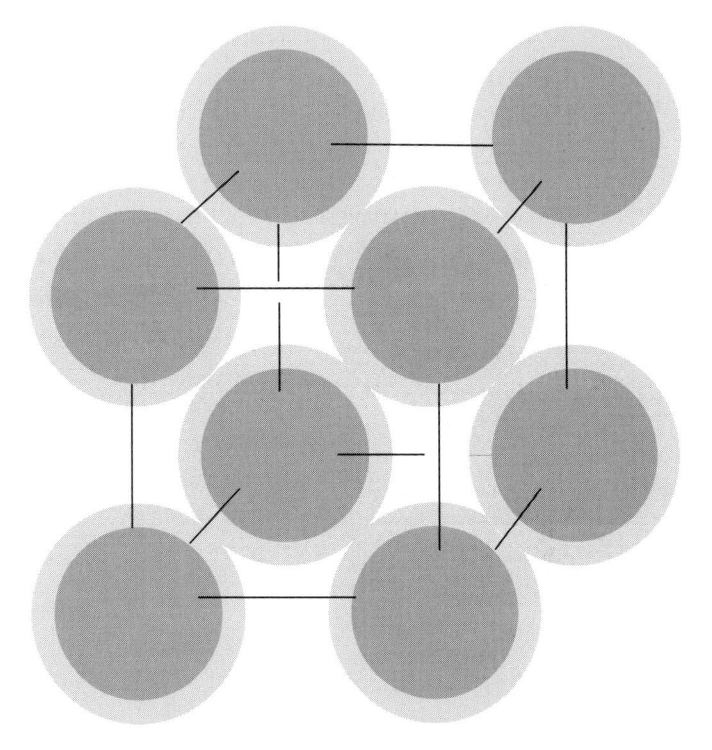

FIGURE V-3. Thermal vibrations of atoms in a lattice. The blurring around each atom (denoted by circles) is meant to suggest how the atom vibrates about, but does not move away from, its mean position. The lines joining the atoms are just to guide the eye.

vibrate to and fro within a small volume around this site. Figure V-3 shows what a unit cell of a cubic lattice might look like in this state. The figure gives you a fair sense of the size of the atom in relation to its distance from the neighbouring atoms. We recall from chapter II that each atom in the crystal feels a force from each of the atoms in its neighbourhood, and the atom sits quietly in its place only because these forces cancel one another owing to the symmetry of the crystal. If I now set just one atom in vibration, this will be transmitted to the other atoms. It is rather like a row of people sitting rather close together on a bench with linked elbows. If one of them begins to rock from side to side, all of them will be doing so too. Note that the atoms do not vibrate independently of one another. The motion of one influences, and is influenced by, the motion of all the others, because there are forces between them: remember that each atom consists of a nucleus and electrons with electric charge. The result is a complicated vibrating motion of all the atoms, and I shall show you in chapter VI how to get a simple picture of this motion. For the present, I just note that there is again an energy associated with this motion, and when the temperature is raised the motion somehow changes so that the energy also increases.

Let me summarize what we have seen in this section. There is a very close connection between the temperature of a piece of matter and the energy of motion of its constituent atoms. While we have considered examples of matter composed of atoms, the connection between temperature and the energy of the constituent particles holds even if the particles are molecules, electrons, ions, and so on. We can go even further: we saw that light has also particle-like properties, and we call these particles photons. The inside of a hot oven is filled not only with hot air but also with infrared radiation which is after all a form of light and is composed of photons. It is as if the oven is filled with a gas of photons, all moving at the speed of light and with an energy distribution that will correspond to the temperature of the oven. We shall see presently that this distribution is different from the one for atoms of helium gas.

QUESTION We spoke of percentages of the total number of atoms in different ranges of speeds for the helium gas, leading to the energy distribution which then connects with the temperature of the gas. The photons are all travelling at the speed of light, so that there can be no distribution of their speeds. So what is the quantity whose distribution is the relevant thing for photons?

ANSWER We recall that the energy E of a photon is related to its frequency f through the Planck constant h:

$$E = hf.$$

So we expect that the energy distribution for photons is to be expressed as the number of photons (and correspondingly their total energy) having frequencies in different ranges.

6 Fermions and bosons

We have so far talked about the atoms, electrons and photons in this chapter as if they were particles. We have not taken into account (other than in mentioning their zero-point energy) their particle-wave duality, which requires that these objects must be described using quantum mechanics. We shall now do this, and discover a result that is basic to much of what follows in the book.

I start by recalling some features of a quantum-mechanical description of a single particle. I shall call it particle A, where A can be an electron, hydrogen atom, etc. It is associated with a wave that is described by some wave function $\psi(\mathbf{r}_A,\alpha_A)$ which has some value at every point $\mathbf{r}_A$ in space. This notation reminds me that the particle A is in a particular quantum state labelled by α_A. As an example, the electron in the hydrogen atom is in a state specified by the values of the quantum numbers n, l, m, and s, and when I say the state α_A, I mean these values of the quantum numbers. For the sake of compactness, I shall later use the single symbol q_A to represent both $\mathbf{r}_A$ and α_A. The square of the amplitude that the wave function has at a given point in space is proportional to the probability of finding the particle near that point.

Now suppose I take two particles labelled A and B that are the same kind: they both are electrons, or hydrogen atoms, for example. The wave function of this system of two particles will certainly be different from that of one particle. It will depend on their positions and quantum numbers, which I shall indicate by the single symbols q_A and q_B. I can then write this wave function as $\phi(q_A, q_B)$. Its amplitude varies as the positions of the particles A and B vary. The square of the amplitude, ϕ^2, is proportional to the probability of finding the two particles near the corresponding positions.

> QUESTION We had some pictures to help us see what some of the wave functions for a particle in a box and for a hydrogen atom looked like. Can we also have a picture of what this new wave function looks like?
>
> ANSWER Alas, not so easily. For one thing, we do not have enough information about the particles, such as whether they are in a box and if so what size and shape of box, etc. We do not even know if both of them are electrons or protons or photons or whatever. If I had all this needed information, I could in principle calculate the wave function ϕ. I would then still have difficulty in making a picture of it, for its value will depend on the positions of each of the two particles, and there is no simple way I could show all of this on a flat sheet of paper.

With such well-nigh total ignorance of the details of the wave function for two particles, one might think that there is little more we can say. Before giving up, we note that there is one point we have not yet made use of: the two particles are identical. If I imagine the particles to be interchanged so that A becomes B and B becomes A, the wave function would change from $\phi(q_A, q_B)$ to $\phi(q_B, q_A)$. But since the two particles are identical, the system will have the same properties after the interchange as before. The simplest of the properties that are related to the wave function is the probability of finding the two particles near any two chosen points. This probability, equal to the square of the wave function, must be unchanged by the interchange of A and B:

$$\phi(q_A, q_B) \text{ multiplied by itself} = \phi(q_B, q_A) \text{ multiplied by itself.}$$

There are two ways of satisfying this condition. One is if

$$\phi(q_A, q_B) = \phi(q_B, q_A),$$

and the other (remembering that a negative number multiplied by itself gives a positive number) is if

$$\phi(q_A, q_B) = -\phi(q_B, q_A).$$

In words: the wave function for a system of two particles must be such that it will, when the two particles are interchanged, either remain exactly the same or at most change its sign (positive or negative) to the opposite sign.

We shall be interested in systems not with just two particles, but with a very large number N, typically 10^{21}, of particles. The wave function for such a system would have the form $\phi(q_1, q_2, \ldots, q_N)$, where each q stands for the position and quantum numbers of one of the particles. Using the same reasoning as I did for two particles, I see that interchanging the two particles at q_1 and q_2 and leaving the others where they were will give me either the same wave function or merely change its sign: either

$$\phi(q_1, q_2, \ldots, q_N) = \phi(q_2, q_1, \ldots, q_N),$$

or

$$\phi(q_1, q_2, \ldots, q_N) = -\phi(q_2, q_1, \ldots, q_N).$$

You will note that in all the foregoing, I have said nothing about the actual form of the wave function, other than that it describes a system of N identical particles. The actual form will depend, for example, on whether the system is the collection of 92 electrons in an atom of uranium or the 10^{21} or so atoms in a container of helium. In any case, the wave function must be one or the other of the two kinds described above. What the particles are – electrons, photons, helium atoms, etc. – determines which of the two kinds is appropriate.

I introduced in chapter II the idea of symmetry: if after I do something to an object it looks exactly the same as before, I say that it is symmetric under what I did to it. In the same spirit I call the wave function that is unchanged when I interchange two particles a *symmetric wave function*, and the one that changes its sign an *antisymmetric wave function*. The operation I carry out in each case is the interchange of two particles. Particles that have symmetric wave functions are called *bosons*, and those which have antisymmetric wave functions are called *fermions*. The names commemorate the physicists Satyendranath Bose and Enrico Fermi who first clarified these ideas. Photons and hydrogen molecules are examples of bosons, electrons and protons of fermions. If I interchange two particles A and B in a system containing identical particles, then I have

$$\text{for bosons: } \phi_s(q_A, q_B) = \phi_s(q_B, q_A),$$
$$\text{and for fermions: } \phi_a(q_A, q_B) = -\phi_a(q_B, q_A).$$

In what I have just written, I have used subscripts s and a to remind me which wave function is *s*ymmetric and which is *a*ntisymmetric. Also, I have not indicated anything about the other particles, which remain undisturbed during the interchange of A and B.

So we have two kinds of particles according to quantum mechanics, fermions and bosons. How do we know which kind a given particle, say an electron, or a proton, or a photon, is? One way is to measure the properties of an assembly of such particles. We shall see in the next section that some of these properties can be strikingly different depending on whether the particles are fermions or bosons. We would then find that electrons and protons are fermions, photons and hydrogen molecules are bosons. This tells us what they are, but not why. The answer to that question is given when we realize that these particles are to be described using not only quantum mechanics but also the Einstein theory of relativity. It turns out then that the spin quantum number of a given particle can only have one of the values 0, $\frac{1}{2}$, 1, $\frac{3}{2}$, 2 and so on. No number lying in between is possible. Particles with integer spin quantum numbers (0, 1, 2,...) are found to have symmetric wave functions and therefore are bosons, and the particles with fractional spin quantum numbers have antisymmetric wave functions and so are fermions. The electron with spin quantum number $\frac{1}{2}$ is therefore a fermion, and the photon with spin quantum number 1 is a boson.

Now whether we have a system of bosons or one of fermions, we have seen that interchanging two particles leaves the system exactly as before, in terms of the probability of finding the particles in specified positions: this after all is what the square of the wave function's amplitude gives us. So one might well ask what difference it makes whether the particles are bosons or fermions. The answer is that it makes a tremendous difference, as we shall see now.

7 Fermions and the Pauli principle

I go back to the case of two identical fermions A and B, and use the symbol q_A for the position vector and quantum numbers of A, and similarly q_B for B. Interchanging the fermions then changes the sign of the wave function:

$$\phi(q_A, q_B) = -\phi(q_B, q_A).$$

I have assumed so far that the fermions have different sets of quantum numbers. For example, if I think of the two electrons in a helium atom, and label them A and B, then their quantum numbers might be as shown in Table V-2.

Table V-2. *Quantum numbers for the two electrons in a helium atom*

electron	quantum number			
	n	l	m	s
A	1	0	0	$\frac{1}{2}$
B	3	2	1	$\frac{1}{2}$

Now consider the special case where the two sets of quantum numbers for the two fermions A and B are identical. I denote this single set by the symbol α. Then the wave function depends on α and the position vectors $\mathbf{r}_A$ and $\mathbf{r}_B$ of A and B, and can be written as $\psi(\mathbf{r}_A, \mathbf{r}_B, \alpha)$. Interchanging A and B leads to the result

$$\psi(\mathbf{r}_B, \mathbf{r}_A, \alpha) = -\psi(\mathbf{r}_A, \mathbf{r}_B, \alpha).$$

I have labelled the fermions A and B just to keep track of their interchange. They are nevertheless identical and have exactly the same quantum numbers, and so interchanging them must leave their wave function unchanged. Yet because they are fermions, the wave function must change its sign. The only way this is possible is if the wave function is equal to zero, for both plus zero and minus zero are just zero. If the wave function is zero, it means that the probability of finding the particles is zero: *I can never find two fermions in a situation where they have all their quantum numbers the same.* A given quantum state, defined by its set of quantum numbers, can be occupied by only one fermion and no more. Fermions are not very friendly types, it would seem. This is exactly the Pauli principle, which we used in describing the atom in chapter IV.

The Pauli principle holds only for fermions. For bosons, interchanging two particles that have the same quantum numbers does not change the sign of the wave function and so it remains the same as before, as is to be expected. This means that in an assembly of bosons, any number of particles can all be in the same quantum state and share the same quantum numbers. In contrast to fermions, which do not seem to like one another's company, bosons are rather gregarious creatures.

8 The quantum states of particles in a box

I want now to develop a quantum description of a very large number N of identical particles. If I find out what quantum numbers to use for this system, then I can see how to assign them to the particles, depending on whether they are fermions or bosons. I assume that the particles are confined inside a three-dimensional box, so that the probability of finding them outside the box is zero. Now you recall from chapter III (page 55) that the analogous condition for a particle in a one-dimensional box led to the wave function taking the form of standing waves with only certain values for the wave number and nothing in between. These wave numbers k_n were the quantum numbers of that system.

In a three-dimensional box, the waves belonging to a particle are in a three-dimensional space, and therefore have wave *vectors* with both magnitude and direction. Again, because the waves must have zero amplitude outside the box and on its walls, we are only allowed standing waves whose wave vectors $\mathbf{k}_n$ can only have particular values of the magnitude k_n and point in particular directions. Each allowed $\mathbf{k}_n$ is then a quantum number (or strictly speaking, a quantum vector) with its own wave function, and the energy E_n of a particle with this wave function is given by

$$E_n = \frac{\hbar^2(k_n)^2}{2m}$$

where k_n is the magnitude of the vector $\mathbf{k}_n$. The wave function for each of these wave vectors will also depend on the spin quantum number s, and I can write it as $\psi(\mathbf{k}_n, s)$. If for example the particles are electrons, then for a given $\mathbf{k}_n$ there are two wave functions, $\psi(\mathbf{k}_n,+\frac{1}{2})$ and $\psi(\mathbf{k}_n,-\frac{1}{2})$ for the two possible values of the spin quantum number for the same value of $\mathbf{k}_n$. How the particles are distributed among these wave functions depends on whether they are fermions or bosons, as we shall see in the next section.

9 Fermion and boson distributions

I shall first consider the case of fermions. I imagine a large number, like 10^{21}, of fermions in the box, and see how they are distributed among the different states

$\psi(\mathbf{k}_n, s)$. I assume to begin with that the box is at the absolute zero of temperature, 0 K, and then see what happens as I heat the box and its contents to higher temperatures.

At 0 K, the fermions must have the lowest possible energy: this is what is meant by the absolute zero of temperature. I show in fig. V-4 a portion of the possible energy

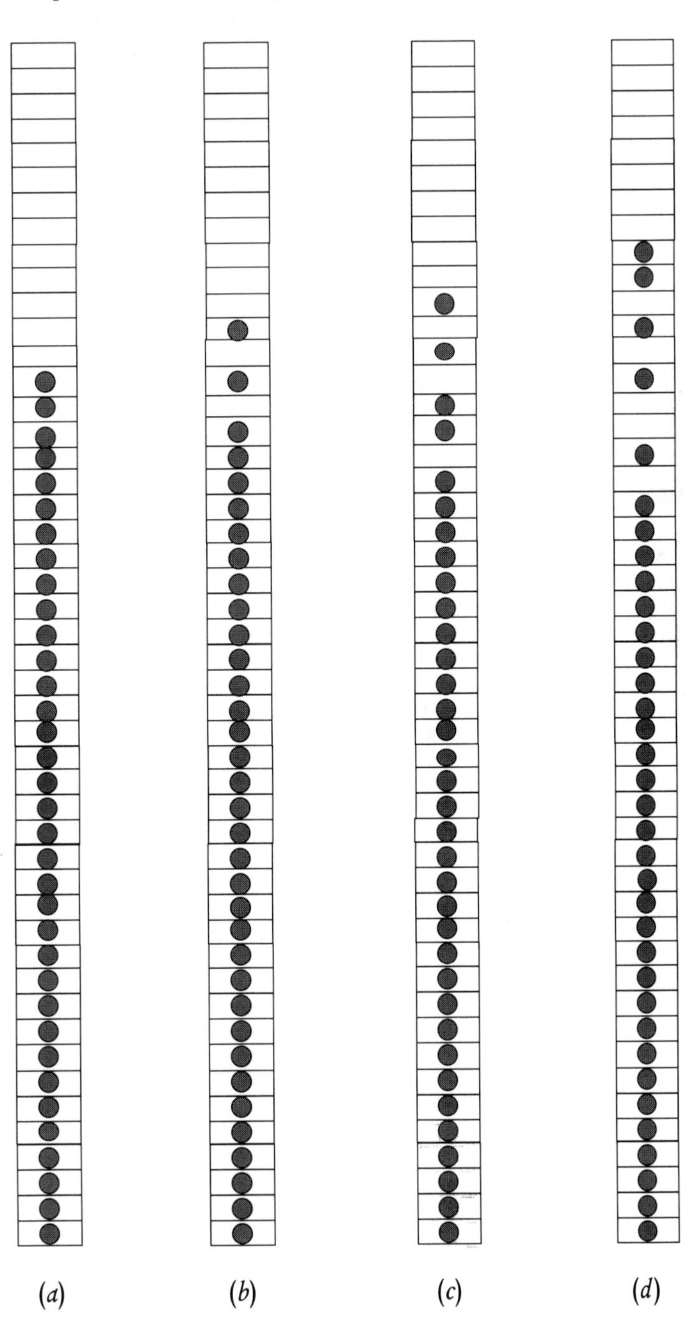

FIGURE V-4. The occupation of successive states of increasing energy by an assembly of fermions, at different temperatures labelled in increasing order from (*a*) to (*d*). Each small rectangle represents a quantum state which is either empty or has one particle (Fermi-Dirac statistics). The energy of the state, allowing for degeneracy, increases as one goes up the stack of rectangles. As the temperature increases, only the top of the distribution is altered, with some fermions moving to quantum states of higher energy and thus leaving empty states behind.

states as boxes stacked in a tower with the lowest energy at the bottom and the energy increasing as I go up the stack. Each of these states has a finite number, certainly very much smaller than 10^{21}, of possible $\mathbf{k}_n$ vectors and corresponding wave functions $\psi(\mathbf{k}_n, s)$. Since no two fermions can have the same $\mathbf{k}_n$ and s, we cannot have all the fermions sitting in the lowest energy state. Rather, they must occupy progressively higher energy states until all the fermions have been accommodated. This situation is shown in fig. V-4, stack on the left labelled (*a*). Every energy state up to a certain highest energy is fully occupied and cannot take any more fermions. This highest energy is called the *Fermi energy*, and it depends on the number of particles, their mass, and the size of the box. Even at the absolute zero of temperature, only one fermion is in each of the states of lowest energy. All the others are in successively higher energy states, one in each state, and are whizzing about at various speeds that can reach quite high values.

QUESTION Does this mean that no two fermions can have the same energy?

ANSWER No. The Pauli principle says that no two fermions can be in the same quantum state. However, quantum states can have degeneracy, meaning that more than one such state can have the same energy. Thus many fermions can have the same energy even though each is in a different quantum state. In a cubic box, for example, there are twelve degenerate states all of the same lowest possible energy.

Now suppose I heat the box and its contents to some temperature above the absolute zero. This means that I add some energy to the fermions, so that they move to nearby states of higher energy. Note that fermions sitting in states well below the Fermi energy cannot go to states of somewhat higher energy, because these states are already fully occupied and will not accept any more fermions, thanks to the Pauli principle. Only those fermions that are at or near the Fermi energy can find empty states at higher energies into which they can move. Fermions have room only at the top. This state of affairs is illustrated in fig. V-4, stacks labelled (*b*), (*c*), and (*d*) for successively higher temperatures. A change in the temperature affects only those fermions with energies near the Fermi energy, which form only a very small fraction of the total number, and leaves the vast majority of the fermions in the same states they occupied at the absolute zero. Electrons are fermions, and many of the properties of solids that we shall be looking at involve changing the energy of the electron system. Thus these properties are mainly determined by just the relatively few electrons with energies near the Fermi energy, while the rest of them are, so to speak, passive spectators.

The situation is quite different with bosons, as illustrated in fig. V-5. The part labelled (*a*) shows the case for 0 K, and (*b*), (*c*) and (*d*) for successively higher temperatures. At each temperature, the box labelled (1) represents the state of lowest energy, and (2), (3) and (4) are three higher energy states. Being bosons, the particles can all get into the lowest energy state, and this is exactly what happens at 0 K. With increasing

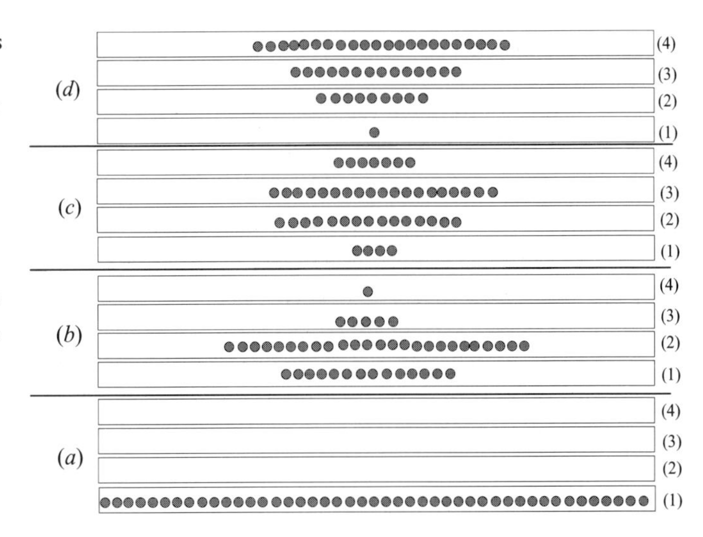

FIGURE V-5. The occupation of states of increasing energy by an assembly of bosons (Bose-Einstein statistics), at different temperatures labelled in increasing order from (*a*) to (*d*). The same four energy states are shown at each temperature. They are labelled (1) to (4) in order of increasing energy. Note the stark contrast to the distribution of fermions shown in fig. V-4, all because the wave function is symmetric for bosons and antisymmetric for fermions.

temperature, more and more of them acquire thermal energy and get into the higher energy states, as shown in parts (*b*), (*c*), and (*d*) of the picture. You can see that the resulting distributions are totally different from those for fermions. It is therefore to be expected that the properties of boson systems will be quite different from the properties of fermion systems.

10 Summary

I introduced in section 1 the idea of a statistical distribution as a convenient way of handling the individual characteristics of the members of a very large assembly. I illustrated it with the example of the distribution by age of the population of a country. Such a distribution contains interesting and useful information about the population that is not so easily accessible from a list of the ages of each of the inhabitants.

In section 2, I considered a volume of gas consisting of a very large number of atoms, something like 10^{21}. These atoms are moving about at all possible speeds, bouncing off one another and off the walls of the container. I showed a picture of their distribution by speed at two temperatures. The distribution changes with a change in temperature, such that the average speed of an atom increases when the temperature increases. I introduced a new way of expressing temperature, in kelvin (K), obtained by adding the number 273 to the temperature in degrees Celsius (°C). Thus a temperature of 20 °C is the same as 293 K. The temperature in kelvin is called the absolute temperature.

I took a closer look, in section 3, at the distributions by speed of atoms in a gas, and found that the average kinetic energy of an atom is proportional to the temperature expressed in kelvin: doubling the temperature of the gas doubles this energy. I thus have a connection between the average energy, which depends on the energy of each of the 10^{21} atoms, and the temperature, which is a property of the gas as a whole. This connection is through the Boltzmann constant k_B. The making of such connections between atomistic and macroscopic aspects of matter is the essence of statistical physics.

I introduced the idea of the absolute zero of temperature in section 4, as a consequence of the average energy being proportional to the absolute temperature of the gas. In practice, one can never reach the absolute zero. Nature is such that the closer one gets to it, the harder it becomes to get any closer.

I saw in section 5 that the temperature of a crystal is related to the energy of vibration of its atoms about their average positions. These thermal vibrations vanish at the absolute zero of temperature, leaving only the zero-point vibrations of the atoms.

I considered the quantum mechanics of an assembly of identical particles in section 6. I found that interchanging any two particles in the assembly must either leave the wave function unchanged (a symmetric wave function), or must change its sign (plus to minus or vice versa) and nothing else (an antisymmetric wave function). Which of the two happens depends on whether the particles are electrons, or photons, or hydrogen molecules, etc. Particles that are described by symmetric wave functions are called bosons, and the other kind are called fermions. For example, photons and hydrogen molecules are bosons, while electrons and protons are fermions.

I showed in section 7 that because fermions have antisymmetric wave functions, they obey the Pauli principle: no two fermions in the assembly can have all their quantum numbers the same. At least one of the quantum numbers must be different for the two fermions.

In section 8 I looked at the quantum mechanics of a particle in a three-dimensional box. This is an extension of the case of a particle in a one-dimensional box that I described in chapter III. I found that the wave vectors and the energies are quantized: they can have only certain values and nothing in between. These wave vectors, along with the spin, can therefore be thought of as the quantum numbers. The quantum states are degenerate: states with different quantum numbers (i.e. wave vectors and spin) can have the same energy.

I took, in section 9, the case of a large number of particles in the box. The way these particles occupy the different quantum states is quite different depending on whether they are fermions or bosons, because of the Pauli principle. I showed with the help of pictures how the different states are occupied at different temperatures in the two cases.

This account of statistical physics, together with the preceding chapters on crystals, quantum mechanics and atomic structure, provides the foundation on which we can now build an understanding of why things are the way they are. As the first step, we shall look at a crystal that is at some temperature and is composed of identical atoms each consisting of a nucleus and its surrounding electrons, to be described using the quantum mechanics and statistical physics which we have now learned. This is the subject of the next chapter.

VI THE QUANTUM MECHANICAL CRYSTAL

1 From isolated atom to crystal

A crystal is an orderly arrangement of atoms that are sitting rather close to each other, as we saw in chapter II. We then looked at the basic ideas of quantum mechanics in chapter III. We used these ideas to get a picture of an isolated atom in terms of its nucleus and especially its electrons in chapter IV. The material things that we deal with in our daily life consist not of single atoms or electrons, but rather of very large numbers (like 10^{21}) of particles. We need a statistical approach to cope with such numbers. We saw in chapter V that quantum mechanics, when used to give such a description of assemblies of particles, divides them into two types: fermions and bosons.

The only property that the crystals of chapter II had was their symmetry. A real crystal has in addition many other properties like being a good (or bad) conductor of electric currents, having a colour, and so on. These properties depend on the particular kind of atoms that make up the crystal. For example, silver and silicon are both crystalline, but differ widely in many of their properties. The only essential difference, other than in the mass, between the atoms of any two elements is in the number of electrons surrounding the nucleus. So we can conclude that this difference must play a key role in determining the specific properties of materials. It is therefore reasonable to see what happens to the electrons when atoms come close together to form a crystal. We proceed to do this using the ideas of both quantum mechanics (because that is the correct description of nature) and statistical physics (because of the large number of particles involved, and also because we shall be interested in effects at different temperatures).

I remind you that the typical size of an atom is a few angstroms (one angstrom = 10^{-8} cm), which is also the typical distance between neighbouring atoms in a crystal. In an isolated atom that is far from all other atoms, each electron feels the influence of the electric charge of the nucleus and of the other electrons in the atom, and nothing else. The result is the set of energy levels and quantum numbers described in chapter IV. When the atoms come together to form a crystal, an electron from a given atom feels the electric charge not only from that atom, but also from neighbouring atoms. It seems reasonable that the wave functions, energy levels and quantum numbers should now be different from what they were in the isolated atom. We shall find out in the next section what these differences are.

2 From energy levels to energy bands

I begin with the isolated atom, and look a bit more closely at how the electrons are distributed around the nucleus. To be specific, I consider an atom of silver, consisting of a nucleus with an electric charge of 47 times $+e$, surrounded by 47 electrons each with a charge $-e$. I use here the symbol e to stand for the magnitude of the charge of the electron. The positive charge of the nucleus is exactly cancelled by the total negative charge of the electrons, so that the total charge of the atom is zero, i.e. it is electrically neutral.

To see how these 47 electrons are distributed around the nucleus in terms of their energy levels, I imagine that I assemble the atom in a series of steps, starting with the bare nucleus. I then add the electrons one at a time, until I have all 47 electrons in place. Each added electron must go into the lowest quantum energy state that is not already occupied by an electron: remember that electrons are fermions. Therefore there can be only one electron in each state with a given set of values for the quantum numbers. Figure VI-1 shows the resulting probability distribution of the electrons round the nucleus. The picture is not drawn to scale; the nucleus at the centre, with its charge of

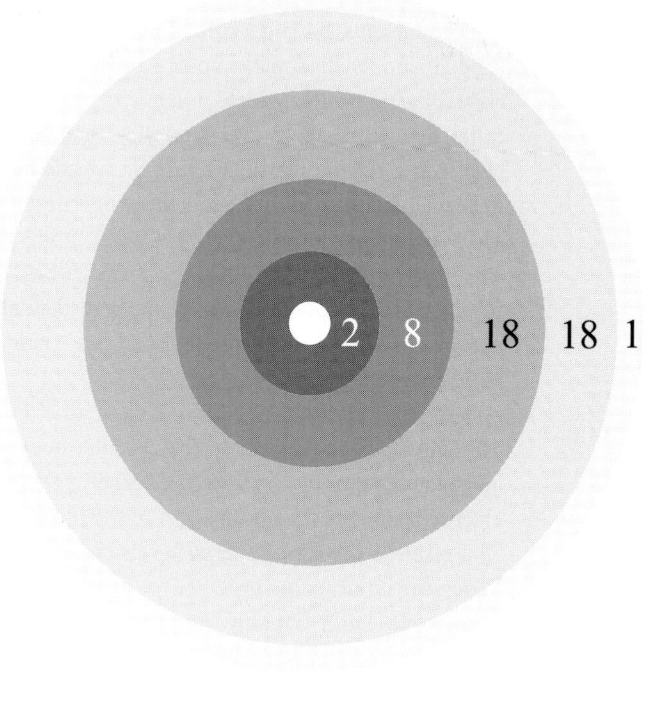

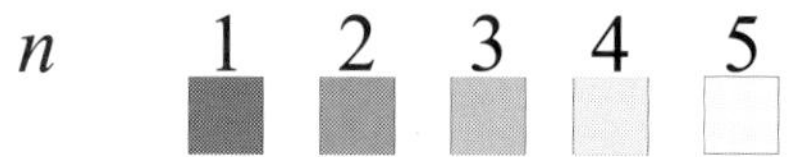

FIGURE VI-1. The silver atom. The white circle at the centre represents the nucleus with a charge of $+47e$. The surrounding shaded regions are where the electrons with different quantum numbers n are most likely to be. The values of n for the different shadings are shown in the key at the bottom. Appearing in each shaded region is the number of electrons in it: two in the region with $n = 1$, for example. The picture is not to scale.

+47e, is very small compared to the size of the atom. The differently shaded regions show where electrons with different quantum numbers *n* are most likely to be, and the number of these electrons is shown for each region. Thus there are two electrons with $n = 1$ very close to the nucleus, and one electron with $n = 5$ near the surface of the atom, and the remaining electrons are distributed in between.

The energy states of the hydrogen atom correspond to the values of the quantum number *n* (p. 72). For a given value of *n*, the different possible values of the other quantum numbers *l*, *m* and *s* gave different wave functions but the same energy: the energy level was degenerate. The situation is different in an atom with several electrons, as happens with silver for example. Here, the energy of a given electron is determined not only by the nucleus, but also by all the other electrons. So the wave functions that correspond to a given value of the quantum number *n* will have slightly different energies for the different sets of values of their quantum numbers *l*, *m*, and *s*.

FIGURE VI-2. A few of the energy levels for an atom with many electrons, such as silver. There are, for example, 18 levels for the quantum number $n = 3$ corresponding to the different possible values of the quantum numbers *l*, *m* and *s*.

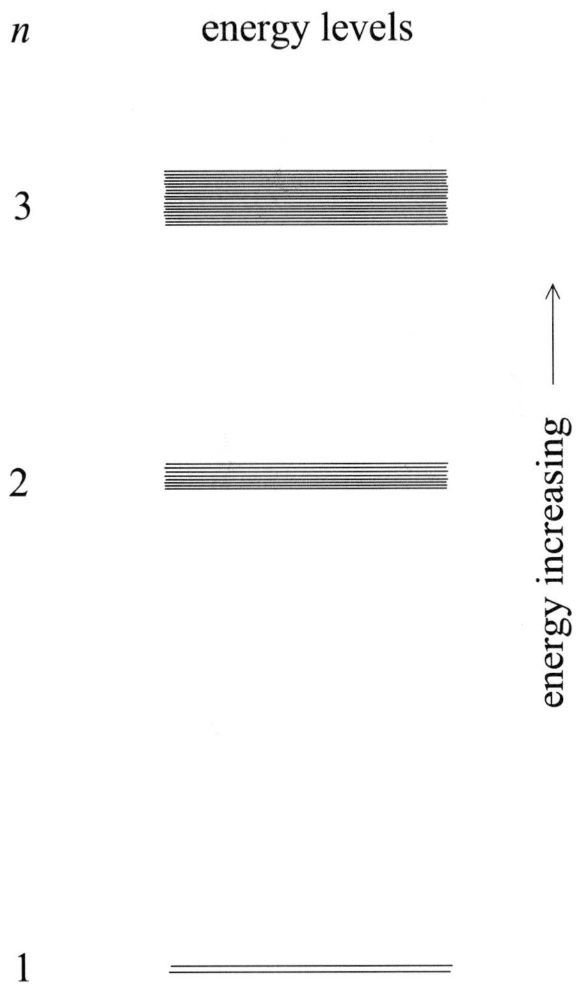

I show the resulting picture of the energy levels in fig. VI-2 for the quantum number *n* equal to 1, 2, and 3. For each of these values of *n*, there is a certain number of levels that are clustered together, and there are gaps in energy between adjacent clusters. Each level corresponds to a different wave function with its set of quantum numbers *n, l, m* and *s*. We know from the Pauli principle that there can be at most one electron described by each of these wave functions with the corresponding energy.

An actual piece of solid, weighing a few grams, consists of a very large number *N* (something like 10^{21}) of atoms sitting very close to one another. We can see what the energy levels for the electrons in such a solid look like by going through the following imagined experiment. We begin by putting the *N* atoms very far apart from one another. Each atom then feels practically no effect from the presence of the other atoms, since the electric force decreases as the distance increases. The energy levels therefore will look practically the same as for the isolated atom (fig. VI-2). But there will now be *N* electrons in each energy level, one electron belonging to each of the *N* atoms. This does not violate the Pauli exclusion principle, because each of these electrons has a wave function belonging to a particular atom with vanishing probability of being near any other atom. This is equivalent to saying that the wave functions of the different atoms do not overlap.

The atoms are now gradually brought closer and closer together, until they finally end up in the positions they have in the actual crystal. The atoms are now crowded close together, and we can no longer neglect the mutual influence of electrons from nearby atoms on one another. Because of this, each of the *N* electrons that was in a given energy level when the atoms were far apart will now have a different energy in the crystal. Each discrete energy level in the isolated atom now becomes a band of energy levels in the crystal, lying very close to each other and numbering as many as the number of atoms in the crystal. This state of affairs is illustrated in fig. VI-3. I show the bands as just grey patches in the picture. If you imagine it hugely magnified, you will see the distinct levels, each capable of accommodating one electron (the Pauli exclusion principle). The electrons in the crystal are accommodated in these energy bands, starting at the lowest energy level in the lowest band, and going progressively higher in energy. You see that the basic idea is the same as what we had with the atom, starting with the nucleus and progressively adding the electrons to the different energy levels. The difference is that the atomic energy levels arise from a single isolated atom, whereas the energy bands for the crystal are due to all the atoms of the crystal taken together.

You notice that the different bands in fig. VI-3 are separated by energy gaps. It happens in some materials that a pair of adjacent (in energy) bands overlap, instead of having a gap between them. The net effect is still that of bands of allowed energies separated by gaps, as in fig. VI-3. An electron can only have an energy that lies in one of these bands but not an energy in any of the gaps. An analogy: it is as if the rules of travel for cars on a highway were such that they could only move at speeds between, say, zero and 20 km/hour, or between 30 and 40 km/hour, and so on. A speed

FIGURE VI-3. The energy levels for different quantum numbers n in the isolated atom (left). The further splitting of the levels for each n (see fig. VI-2) is not shown. In the crystal (right), there are bands of energy levels separated by energy gaps. Each of these bands has at least as many energy levels as there are atoms in the crystal.

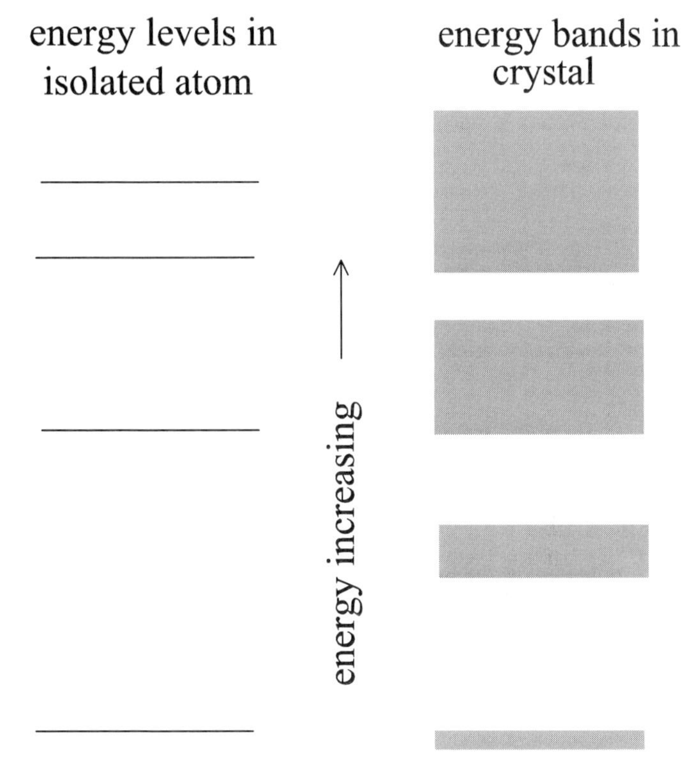

between 20 and 30 km/hour is just not allowed. We would then talk about bands of allowed speeds and bands of prohibited speeds. The corresponding rules for electrons in a crystal are those of quantum mechanics. They say that electrons can only have energies in certain bands, and not in between: we have allowed and prohibited bands of energy. This is a fundamental property of electrons in a crystal, and leads to an explanation of the difference between insulators which cannot carry an electric current, and metals which can, as we shall see in the next section.

3 Insulators and metals

We now populate the energy bands with the electrons in the crystal, beginning with the lowest energy in the lowest band, and going to progressively higher bands until all the electrons have been accommodated. The resulting picture is shown in fig. VI-4. Depending on the total number of electrons per atom, which in turn depends on which chemical element makes up the crystal, the highest band that has electrons will either be fully occupied, or only partially occupied. The two situations are illustrated in the left and right parts of the figure respectively. When the highest band with electrons in it is fully occupied, as on the left of the figure, we get an insulator. When it is only

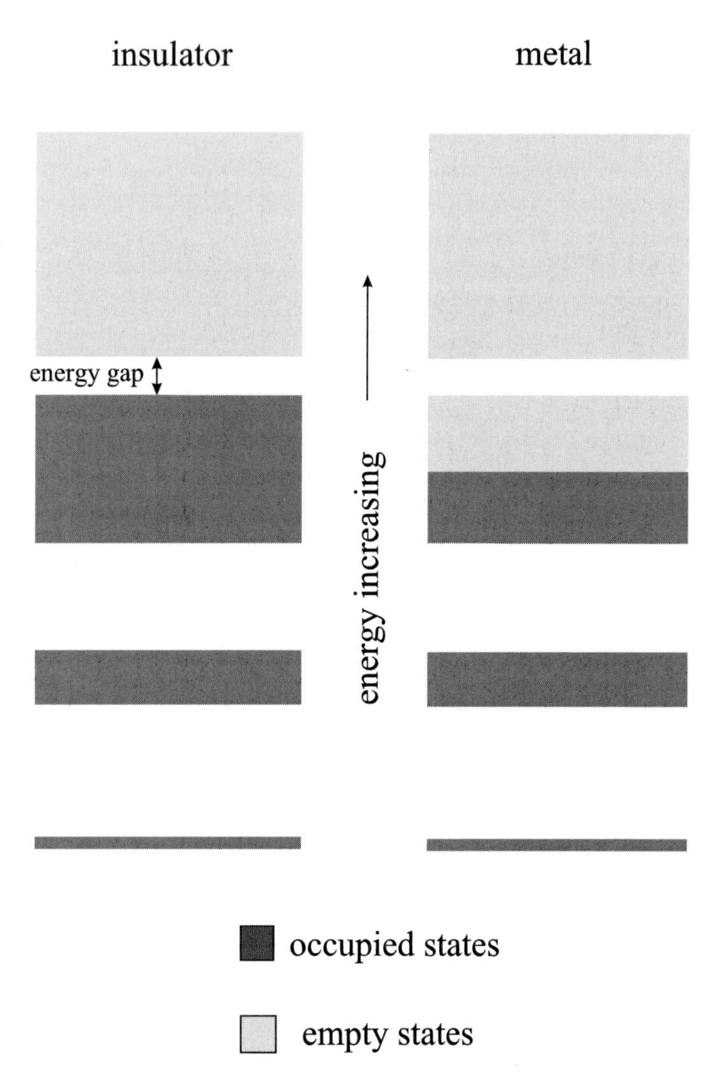

FIGURE VI-4. Energy bands showing occupation by electrons, in an insulator (left) and a metal (right). In the insulator, the band with all its states empty (shown at the top), i.e. without electrons, is separated by an energy gap from the next lower band, in which all states are occupied by electrons. In the metal, the band of highest energy with electrons in it is only partly occupied and has empty states left.

partially occupied, we get a metal. Electrons in the highest energy states in a metal can absorb extremely small amounts of energy because they have suitable empty states to which they can go. Electrons in an insulator, on the other hand, can only absorb energy equal at least to the gap energy, and nothing less. Figure VI-5 illustrates the differing responses of an insulator and a metal to small additions of energy due to a rise in temperature or an applied voltage, for example.

I have described so far how the electrons are distributed in the energy bands. One can also think of how the electrons are distributed in space in the crystal itself. For this I shall take the example of silver as a typical metal, and common salt and silicon as typical insulators. In these as in all other solids, only a few electrons which are farthest from the nucleus are affected significantly; all the other electrons are nearly in the same

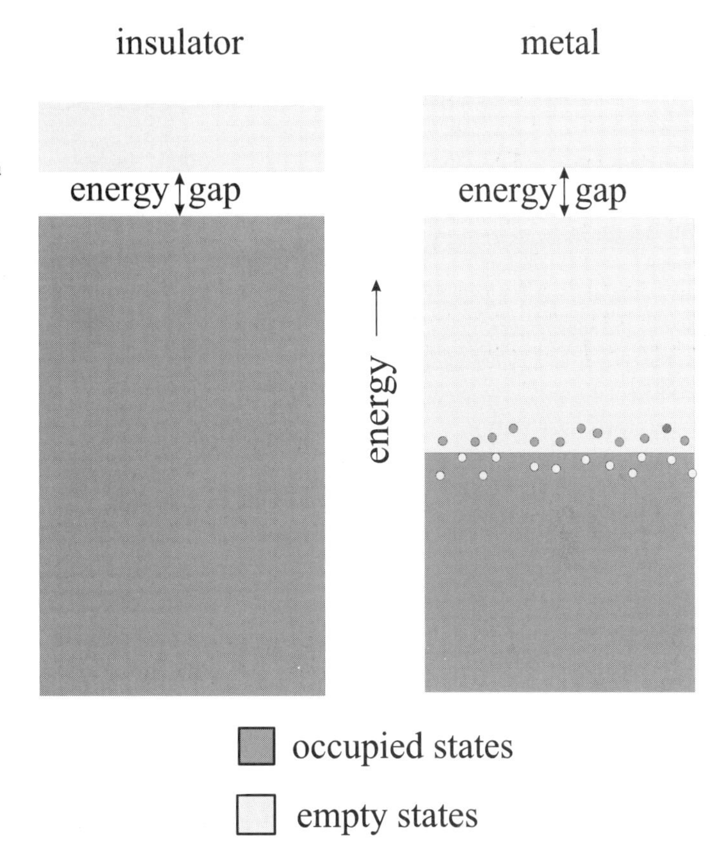

FIGURE VI-5. The effect of a small energy input like an applied voltage or a rise in temperature on an insulator (left) and a metal (right). In the insulator, the electrons are in a full band, there are no empty states nearby in energy into which they could go in response to the energy input, which consequently has no effect on the electrons. In the metal, on the other hand, the electrons respond by changing energy and moving to states nearby which were previously empty (dark circles), leaving behind empty states (light circles).

positions with respect to the nucleus as in the isolated atom. This is to be expected, for the outermost electrons of a given atom have the weakest binding to the atom, and will therefore be influenced by neighbouring atoms most strongly.

I show in fig. VI-6 what a cross-section through a plane of atoms in a crystal of silver looks like, and for comparison a section through an isolated silver atom. The picture is not to scale, but shows the features of interest. The different shaded regions indicate where the electrons with different quantum numbers n are most likely to be found. The electrons with quantum numbers from 1 to 4 are little affected by being in the crystal. The electrons with $n = 5$, however, have parted company from their parent atoms and can be found anywhere in the crystal with equal probability. Each atom is now an ion with an electric charge $+e$, and the crystal is a lattice of these ions immersed in a sea of electrons.

QUESTION Why is the charge on the ion equal to $+e$?

ANSWER The atom has a charge $+47e$ on the nucleus surrounded by 47 electrons each with a charge $-e$, so that its total charge is zero. The outermost electron has wandered off in the metal, leaving behind an ion with a total charge of $+47e-46e$, or just $+e$.

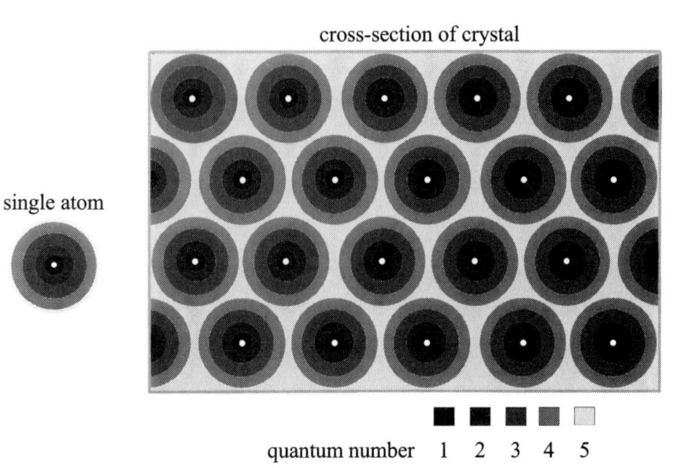

FIGURE VI-6. A cross-section through a plane of atoms in a crystal of silver (right), and through an isolated silver atom (left). The electrons in the inner shells with $n = 1$ to 4 are essentially undisturbed by being in the crystal. The electron in the shell with quantum number $n = 5$ in the atom is set free in the crystal, and has nearly the same probability of being anywhere in the crystal (as indicated by the uniform shading of the space between the ions in the picture on the right).

If the silver crystal is composed of N atoms, then there are N electrons roaming around in it, not attached to the ions. These are the electrons which flow through the lattice of ions when a voltage is applied, causing an electric current. If you think of the current as a stream of particles (electrons) flowing through a dense forest (the lattice of ions), you might expect the electrons to be constantly bounced back by the ions. It would then be difficult to maintain any sort of a current going. Metals nevertheless are very good conductors for an electric current. The explanation comes from quantum mechanics. Electrons behave like waves, and waves, unlike particles, can go past obstacles in their path rather easily. The music we hear from a loudspeaker is little affected by the fabric cover in front of it. On the other hand, dust particles from the outside cannot easily get through the fabric to the loudspeaker cone. The fabric lets waves get through but hinders the passage of particles. There is still another puzzle: we find that metals become even better conductors of electricity as we make them colder. How this comes about is one of the topics in chapter X.

Table VI-1. *Arrangement of electrons in sodium and chlorine atoms*

quantum number n	number of electrons in sodium atom	number of electrons in chlorine atom
1	2	2
2	8	8
3	1	7

The second example I take is common salt, or sodium chloride as it is known to chemists. It consists of equal numbers of sodium and chlorine atoms, arranged in a cubic lattice whose unit cell is shown in fig. II-14. The arrangement of the 11 electrons around the sodium nucleus and the 17 electrons around the chlorine nucleus, when the two atoms are far apart, is shown in table VI-1. The upper half of fig. VI-7 shows the probability distribution of these electrons around their respective nuclei. Each atom is electrically neutral, since the positive charge of the nucleus is exactly cancelled by the total negative charge of the electrons. Further, the shells with n equal to 1 and 2 in both atoms are full, i.e. they cannot accept any more electrons (see p. 78). The shell with $n = 3$ has 18 quantum states and so can accept more electrons, at least as far as the Pauli exclusion principle is concerned. When the atoms come close together to form a crystal, the electron with $n = 3$ on the sodium atom finds that it is happier, i.e. the energy is lower, when it abandons its parent atom and moves into the $n = 3$ shell on a neighbouring chlorine atom. A case of alienation of affections, one might say. But with a difference: instead of hating, i.e. repelling each other, the two ions resulting from the electron's move are positively (sodium) and negatively (chlorine) charged and therefore are pulled towards each other, as shown in the lower half of fig. VI-7. It is this attractive force which holds the crystal together. You can get an idea of how this works from fig. VI-8, showing a plane of atoms in the crystal with alternating sodium and chlorine ions.

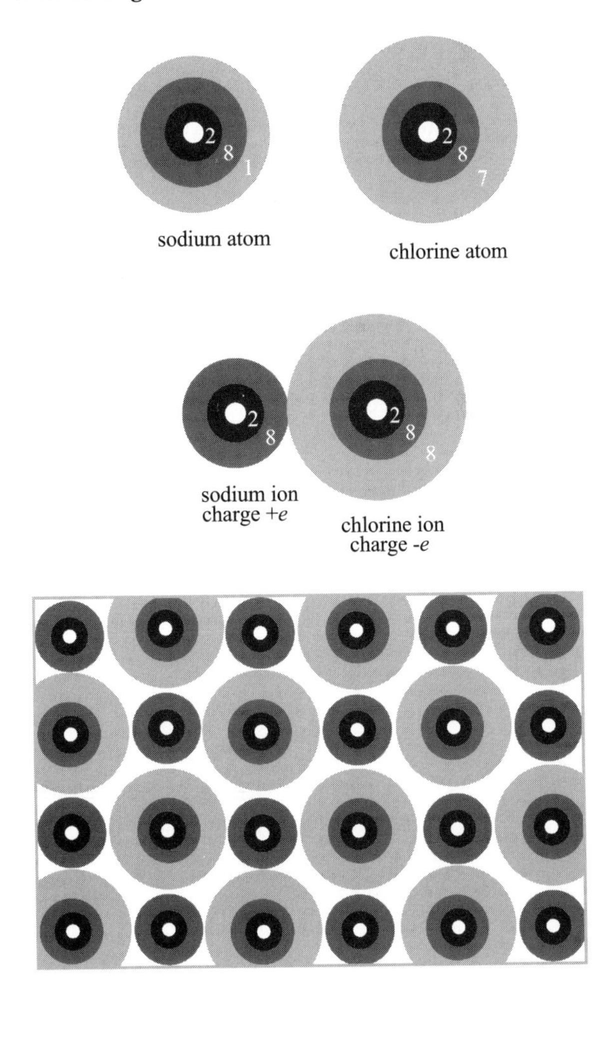

FIGURE VI-7. Isolated atoms of sodium and chlorine (top), and the respective ions (bottom), where the electron from the outer shell of sodium has moved to the outer shell of chlorine. The resulting ions have opposite electric charges and are therefore attracted towards each other. The numbers of electrons in each shell are shown. The small circle at the centre is the nucleus. The figure is not to scale.

FIGURE VI-8. Cross-section through a plane of ions in a crystal of sodium chloride (common salt). The larger ions are chlorine, and are negatively charged. The sodium (smaller) ions are positively charged. The attraction between opposite charges holds the crystal together.

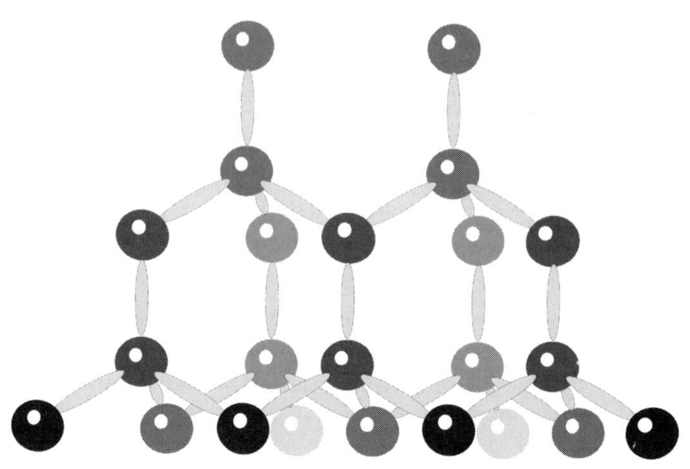

FIGURE VI-9. A small portion of the silicon lattice. Atoms in different planes parallel to the paper are shaded differently, dark to light in going from front to back. Each atom is connected to four neighbouring atoms by what look like cucumbers. Each 'cucumber' represents the region where two electrons, one from each atom, spend most of their time and, together with their adjoining ions, form the bonds which hold the crystal together.

My third and last example of how the outer electrons of atoms are distributed in a solid is silicon. The isolated atom of silicon has a total of 14 electrons around the nucleus. There are two electrons with $n = 1$, eight with $n = 2$ and four with $n = 3$. Again, as with sodium and chlorine, the first two shells are full. When the silicon atoms come together and form a crystal, it has a face-centred cubic lattice, and each lattice point has two atoms associated with it. A small portion of the resulting crystal is shown in fig. VI-9. Pairs of neighbouring atoms are connected by what look like cucumbers. Each of these shows the volume in which two electrons are most likely to be, one coming from each atom in the pair and helping to bind them together. You recall that the outermost electron shell of the silicon atom with $n = 3$ has four electrons. In the crystal, the atom contributes each of these electrons to one of the bonds to the four nearest atoms.

I have described the crystal structure of silicon in some detail because the material is an essential part of electronic gadgets like computers, television, telephone systems and so on, and is therefore an important element in today's life. Another material that is also regarded as very important by some people, namely diamond, has exactly the same structure as silicon, except that the atoms are carbon.

4 Electron waves in a crystal

The free electrons that have left their atoms and are wandering about in the crystal are sometimes called *conduction electrons*, because they are the ones responsible for conducting electric currents in the metal. We would expect their wave functions to be quite different from those in the free atom. The same should also be true for the wave functions of the outer electrons in insulators, since they are the ones that are most influenced by neighbouring atoms.

You recall that we considered in chapters III and V electrons moving about in a box, and saw how they are like waves with different wave vectors **k**. Each of those waves

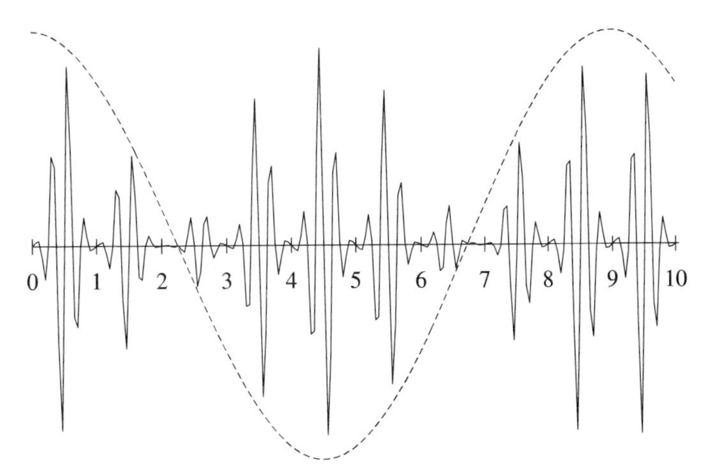

FIGURE VI-10. What the wave looks like for a free electron in a region of a box (broken curve) and for an electron in the same region of a crystal that has the same size and shape (continuous curve). There is a row of atoms (not shown), one in each of the segments between the points labelled 0 and 1, 1 and 2, ... The wave in the crystal, called a Bloch wave, has the same shape at each atom, but its overall amplitude is like that of a free electron wave.

belonged to a different wave function, which could be associated with two electrons with spin up and spin down respectively, both having the same energy.

What do the waves look like for the electrons that were in the outer shells of the atoms in a crystal? The answer is illustrated in fig. VI-10. The continuous curve shows a typical wave for such an electron in a piece of crystalline solid as one goes along a line of atoms in some region inside it. The broken curve shows a wave for a free electron in the same region of a box that has the same shape and size as the solid. The wave for the electron in the crystal has a pattern of wiggles at each atom, but its overall amplitude changes from one atom to the next like that for a free electron wave. One would expect this kind of picture, if one recalls that the wave amplitude at any point tells us about the probability of finding the electron in the neighbourhood of that point. The probability of finding the conduction electron inside the ion is strongly influenced by the other electrons and the nucleus of the ion itself and leads to those wiggles. Since the electron can move from atom to atom, it also has the aspect of a free electron wave. It is rather like some folk dance in which there are a number of identical groups of dancers at different places on the floor. These are the electrons in the ions. The outer electrons are like individual dancers who spend a little time with each group before moving on to the next group. These waves of the electrons in a crystal are called Bloch waves, after the physicist Felix Bloch who first thought of them. Such electrons are referred to as Bloch electrons, to distinguish them from free electrons in a box.

For the free electron in a box, we saw in chapter III that there is a relation between the energy E and the wave number k,

$$E = \frac{\hbar^2 k^2}{2m}.$$

For the electrons in the solid, I would expect the variation of E as $\mathbf{k}$ varies to be a little more complicated, because there are all those ions influencing the electrons. I show in fig. VI-11 how this variation, for fixed direction and changing magnitude of the vector $\mathbf{k}$, might look for two adjacent energy bands with a gap between them. As the wave number increases, the energy in the lower band increases and the energy in the upper band decreases. In this figure, each of the little squares, like the one labelled P, represents a possible state with its specified wave number and energy. For the state P, I can denote the values of wave number and energy by k[P] and E[P] respectively, as read off on the two scales. There can be a maximum of two electrons, with opposite spins, in each of these states (Pauli exclusion principle). I show a small number of states in each band to avoid cluttering the picture. The actual number is comparable to the number of atoms in the crystal, about 10^{21}.

Figure VI-11 gives us more information about the energy bands than is shown in fig. VI-3. We see that each band is made up of closely spaced states, each with its wave function, wave number k, and energy E. There is no state with energy lying in the gap between the two bands. If a given electron changes from one state to another, the

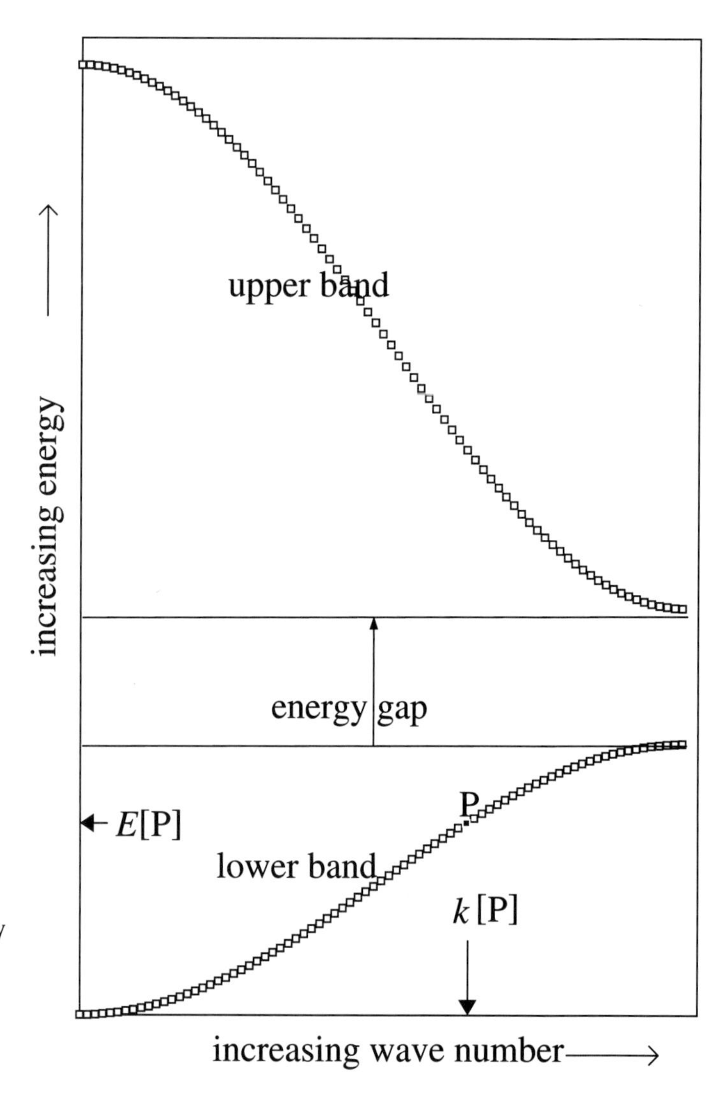

FIGURE VI-11. Two neighbouring energy bands, showing the discrete quantum states (small squares) of which P is an example, with its specific wave number k[P] and energy E[P]. This and the following pictures of energy bands are qualitative, and therefore no numbers appear along the axes.

figure tells us how much the corresponding changes are in the wave number and energy of the electron. An electron can go from one state to another within a band with very small change of energy, but to go from one band to the other must involve a change of energy equal at least to the energy gap.

The energy in a given band may either increase or decrease as the wave number increases. In fig. VI-11, I have arbitrarily chosen the lower band to be one way and the upper, the other. Both types occur among bands in solids, and indeed one can have in the same band both kinds of dependence of the energy on the wave number.

QUESTION With electrons in a box, the energy always increased when the wave number increased. Why is it sometimes different with electrons in the crystal?

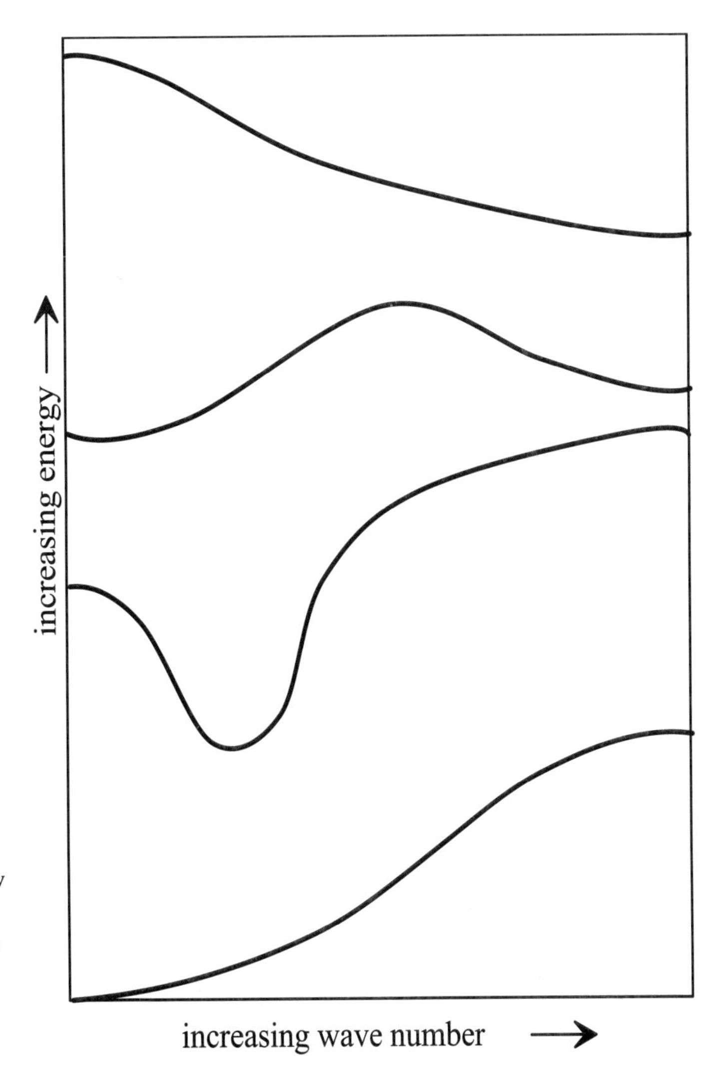

FIGURE VI-12. Some of the different possible types of curves showing how energy varies with wave number for electrons in a crystal. Such curves are said to represent dispersion relations for the electrons, and are called dispersion curves.

ANSWER You see from fig. VI-10 that the electron wave in a crystal is heavily disturbed near the ions, while in the box it looks quite smooth everywhere. When the energies for such wave functions are calculated, it turns out that they sometimes increase, and sometimes decrease, as the wave number increases.

I show in fig. VI-12 a few of the possible shapes for bands. In this figure, I have not shown the little squares representing the separate states in each band as in fig. VI-11, but have joined them as smooth lines. For an actual crystal, there is one of these graphs of energy E versus wave number k for each direction that the wave vector points in the crystal, looking like so many strands of spaghetti in a bowl. Such curves are called *dispersion curves.* A collection of all such pictures for a given solid is called the *electronic*

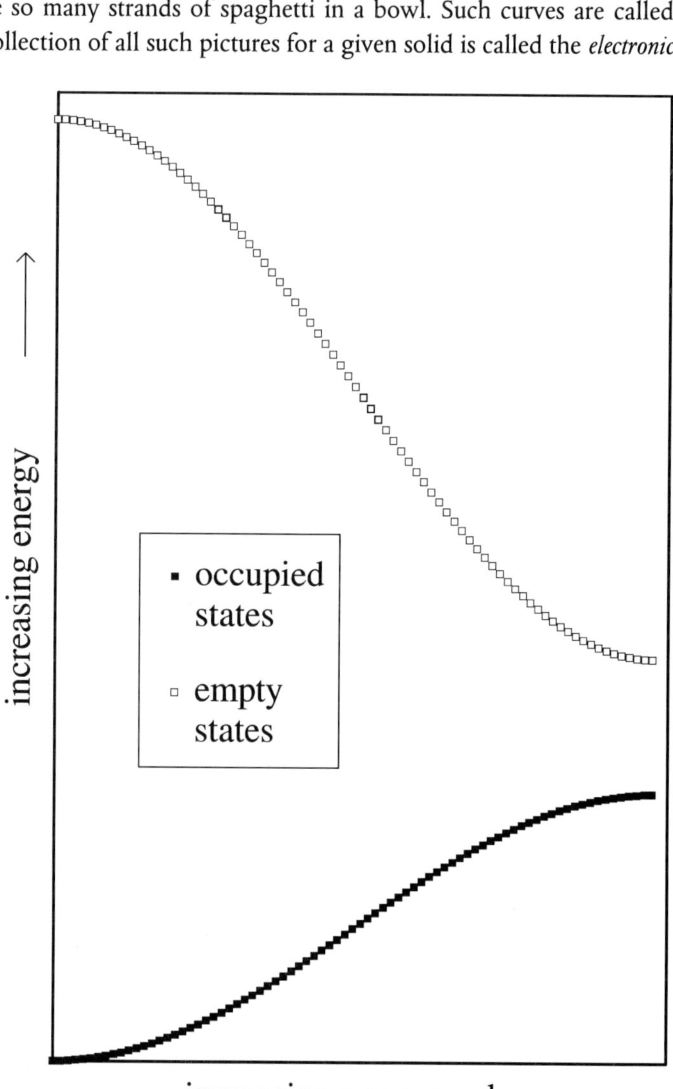

FIGURE VI-13. Energy bands in an insulator (or pure semiconductor) at the absolute zero of temperature. All the states in the lower band are occupied by electrons (filled squares), and there are no electrons in the states in the upper band (empty squares).

band structure, or sometimes simply the *band structure*, of the solid. The crystal structure and the particular element or elements making up the solid determine the actual band structure. It is used to deduce many of the properties of the solid, and we shall next see an example of this.

The electrons in the crystal occupy the states in the energy bands, starting with the lowest energy, until all the electrons have been accommodated. Each quantum state of given energy takes two electrons with opposite spin. Then, depending on how many electrons there are altogether, the last band containing electrons will be either exactly full, or still have unoccupied states. These two possibilities are shown in figs. VI-13 and VI-14 respectively.

The situation in fig. VI-13 is that of an insulator at the absolute zero of temperature. The lower band is completely full, the upper band is quite empty, and there is an energy gap between the two. To set up an electric current means that some electrons have to change to states of slightly higher energy. They cannot do this since the lower band has no empty states for the electrons to move into, and the empty states in the upper band are too far away in energy to be accessible to the electrons. As a result, applying

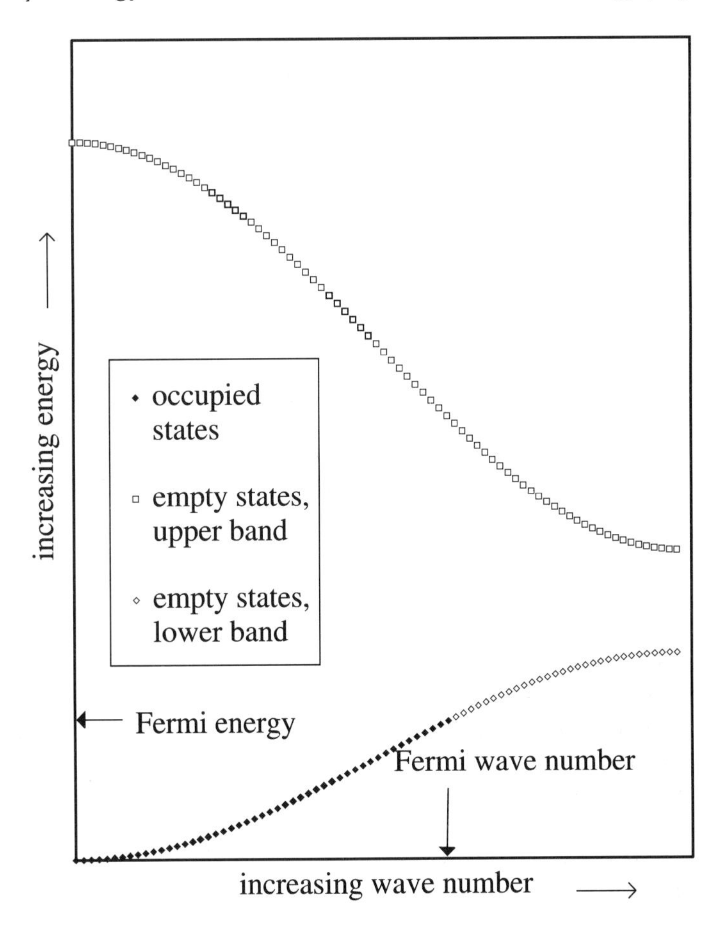

FIGURE VI-14. Energy bands in a metal at the absolute zero of temperature. All the states up to a maximum wave number (the Fermi wave number) in the lower band are occupied by electrons, the remaining states are empty. The energy of this highest occupied state is the Fermi energy.

an electric voltage to the crystal produces no electric current: we have an insulator.

Figure VI-14 shows how the bands are occupied in a metal. The highest band that has electrons still has unoccupied states left. All the lower bands (which are not shown in the picture) are completely occupied by electrons, and the next higher band (shown) is empty. Now if an electric voltage is applied to this crystal, electrons in states below the one that is marked 'Fermi wave number' (more about this in a moment) have empty states to move to, and so an electric current results.

The electron waves can be travelling in all directions in the space of the crystal, and one should properly speak about their wave vectors. If I look at the lower band in fig. VI-14 and think about it in terms of wave vectors, then I would say that all states with wave vectors up to a certain maximum length in that direction are occupied by electrons, and states with larger wave vectors are empty. This largest wave vector is called the *Fermi wave vector*, after the physicist Enrico Fermi. There is such a vector for each direction in the crystal, though their lengths may not all be the same; after all, the crystal looks different in different directions, and so do the electron waves. The magnitude of this vector is the Fermi wave number.

Now imagine a point in space from which radiate arrows in all directions, each being the Fermi vector in its direction. The result will be a hedgehog-like object, and the surface that encloses it, touching the tip of each arrow, is the *Fermi surface.* Figure VI-15 shows what a cross-section through a Fermi surface and four typical Fermi wave vectors might look like. All the quantum states whose wave vectors lie inside the Fermi surface are occupied by electrons, and all those outside are empty. The energy for each Fermi wave vector is the same, and is called the Fermi energy.

QUESTION Why should the energy for all the Fermi wave vectors be the same?

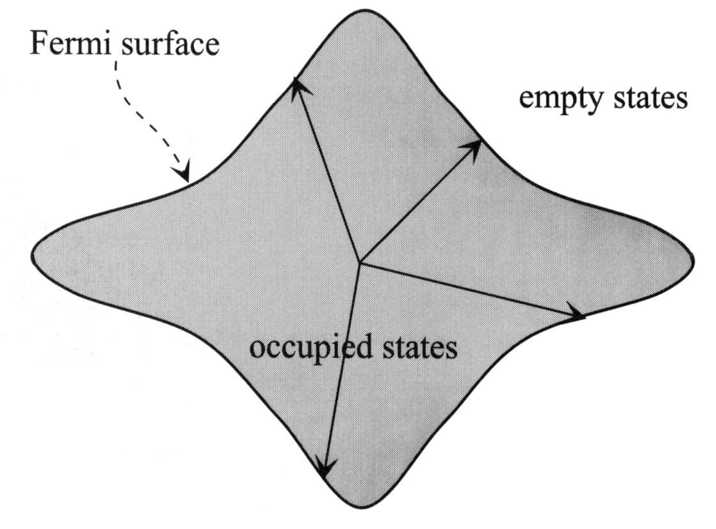

FIGURE VI-15. Cross-section of a Fermi surface, indicated by the closed contour, at the absolute zero of temperature. All the states inside the surface are occupied by electrons, all the states outside are empty. Each state is identified by its wave vector and energy. Four typical Fermi wave vectors are shown, each going from the centre to the Fermi surface. Though they point in different directions and have different magnitudes, they represent states which have the same energy, namely the Fermi energy.

ANSWER Suppose they all had different energies. Then there would be a shifting of electrons from state to state until they all had the same energy. The stable situation is where each electron is in the lowest energy state possible. This is achieved by filling all the energy states of different wave vectors with electrons up to the same energy, for then no electron has an empty state of lower energy to which it can go.

QUESTION What does the Fermi surface look like for electrons filling a box?

ANSWER Since there is no crystalline lattice whose anisotropy would lead to different properties in different directions, the length of the Fermi vector must be the same in all directions. So the Fermi surface will be a sphere.

A metal like silver has tens of electrons per atom. Most of these electrons in the metal are in completely filled energy bands, and play no part in carrying a current. It is only the relatively few electrons in the states of highest energies, with nearby empty states, that carry the current. Put differently: only those electrons that are in states very close to the Fermi surface take part in electrical conduction. All the other electrons are, in a sense, just passive observers of the scene. This is true not only for electrical conduction, but also for most other properties of metals, as we shall see. So when I think of a metal, it helps not only to think of the regular crystalline arrangement of the atoms in it, but also of the Fermi surface associated with the electrons in it. It is in a space in which lengths are measured in wave numbers: the units are reciprocal centimetres (cm^{-1}), and not the usual centimetres you find on a tape measure.

QUESTION Can one talk about a Fermi surface for an insulator also?

ANSWER It is true that in an insulator also there is, in each direction in the crystal, a wave vector of maximum length for the highest occupied state, as one can see in fig. VI-13. I can construct a surface using all these wave vectors in exactly the same way that I got a Fermi surface for a metal. The resulting surface, called a *Brillouin zone* after the physicist Leon Brillouin, has nothing to do with the electrons at all, but is solely determined by how the atoms are arranged in the lattice. This comes about essentially because (a) the electron waves are Bloch waves, (b) the bands are either fully occupied or empty, and (c) they satisfy the Pauli exclusion principle. I must let it go at that; to explain it further will take us far afield. This idea is not useful for an insulator since we are concerned with its electronic properties, and the Brillouin zone just represents the crystal lattice and not the electronic structure. The pictures of energy bands, like those in figs.VI-12, are on the other hand more meaningful for insulators.

5 Semiconductors

You have probably come across the word *semiconductor* in newspapers and elsewhere, and may wonder what it has to do with the metals and insulators that we have been talking about here. A semiconductor, as the name implies, lies somewhere between metals and insulators in its ability to carry an electric current. It is really an insulator to start with, but something happens along the way so that it ends up with a few electrons in states that have empty states very close by in energy, and it is now able to carry an electric current. What happens to it is that it is at some temperature above the absolute zero, or is not pure but has some impurities in it, or is both.

Take the case of temperature first. We saw in chapter V that at any temperature the thermal energy is distributed among the particles over a wide range, all the way from very low to very high energies. So with the insulator at some temperature, there will always be some electrons whose energies are larger than the gap energy. These electrons can therefore move into empty states in the upper band and leave empty states behind in the lower band, as shown in fig. VI-16. Notice in this figure that there are now empty states close to states with electrons in both the upper and the lower bands, so that current can be carried by electrons in both bands. There is a difference, however: the upper band has a few electrons sitting in an otherwise empty band, while the lower band has a few empty states in an otherwise filled band. It turns out that the current in the lower band is exactly the same as what one would have in a similar band containing a new kind of particle called a *hole*, equal in number to the number of empty states in the original band.

The hole is like an anti-electron: each of its properties (like energy, wave vector or spin) is the negative of that of the electron that previously was in that quantum state. When I work out how a hole moves when I apply a voltage, I find that it behaves like an electron, except that it has a positive charge $+e$. This means that if the electron moves to the right because of the voltage, the hole moves to the left. Since an electric current is just a moving stream of charged particles, the upper band gives a current of electrons, and the lower band gives a current of holes. Because their charges as well as the voltage-induced motion are opposite, the two currents are in the same direction and add to each other. The physical reality is that the current is carried by electrons in both bands. The concept of a hole is only a way of describing the properties of the lower band with some empty states. It avoids having to worry about what the huge number of electrons do, and instead focuses on the few holes alone.

This picture of a hole applies also in a metal to the Fermi surface with an empty state left behind by an electron, when it acquires some energy and moves to a state outside the Fermi surface. The energy needed to move an electron to another state must at least be equal to the energy gap in a semiconductor, but can be almost negligibly small in a metal. It is this difference that essentially distinguishes a semiconductor from a metal, and ultimately makes possible the use of semiconductors in transistors, chips, and all the other wonders of the electronic age.

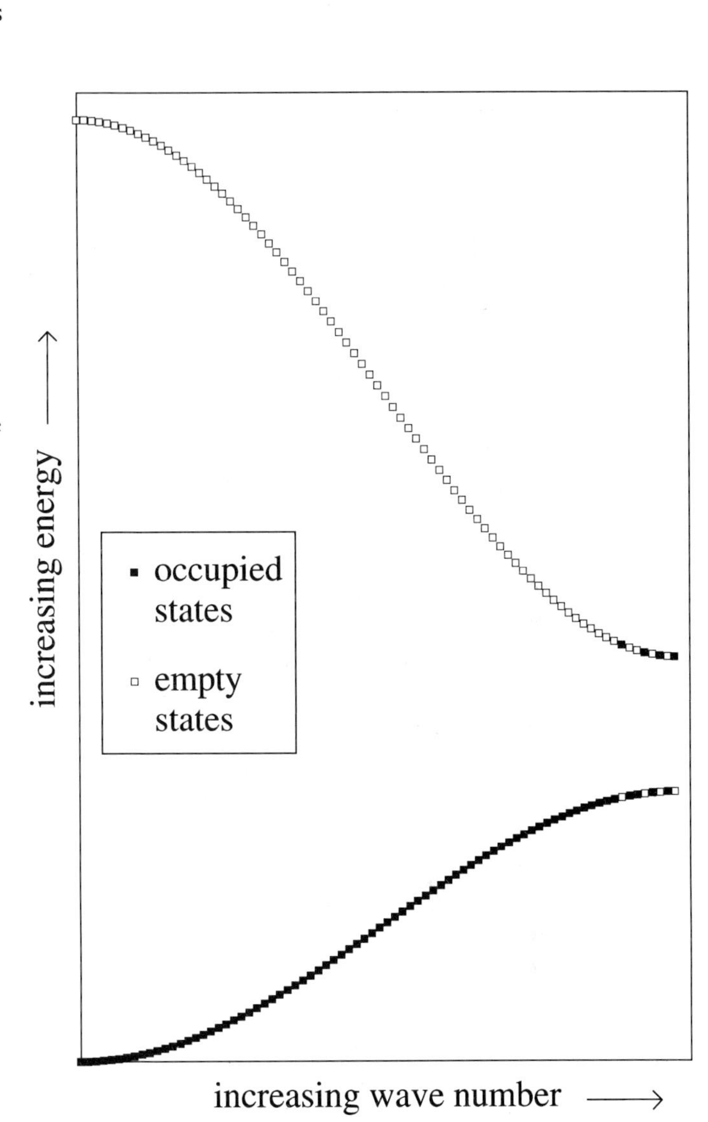

FIGURE VI-16. An insulator (or pure semiconductor) at some temperature above the absolute zero. Some of the electrons in the lower band have acquired enough thermal energy to cross the energy gap and go into states in the upper band, leaving empty states behind in the lower band. Now it is possible to set up an electric current, which is carried by electrons in both bands, because there are empty states next to occupied states in both. The current in the lower band can be thought of as being carried by positively charged particles called holes.

An insulator can change to a semiconductor even without its temperature being raised, when it contains atoms of certain other elements. We saw how in silicon the four outermost electrons on each atom form the bonds with neighbouring atoms. There is an energy gap between the fully occupied band (which is where these four electrons are) and the next higher band, which is empty of electrons. Now suppose we melt some silicon, add a pinch of phosphorus, and then make a crystal out of the mixture. It will be a crystal of silicon *doped* with phosphorus, as the people who make and use such things call it. There are phosphorus atoms in some places in the crystal where there would have been silicon atoms, but otherwise the crystal looks the same as one of pure silicon. The phosphorus atom is only slightly different from the silicon atom: the charge of the nucleus is $15e$ instead of $14e$, and there are five electrons in the outermost shell, instead of four as in silicon. Four out of these five electrons again are used to form the bonds to neighbouring atoms and completely fill up an energy band, just as in pure silicon. But now we have an extra electron from each phosphorus atom which must find a home somewhere. The only place for it to go is the next energy band, which was empty in pure silicon. This is illustrated in fig. VI-17. The electrons in states in the upper band have empty states very close by in energy, and can therefore carry a current

If I dope silicon with a bit of aluminium instead of with phosphorus, the upper band remains empty but now empty states appear in the lower band. This is because the aluminium atom has one electron less than the silicon atom in its outer shell. Some of the bonds therefore have only one electron instead of two, leaving empty states in the lower band that behave as holes.

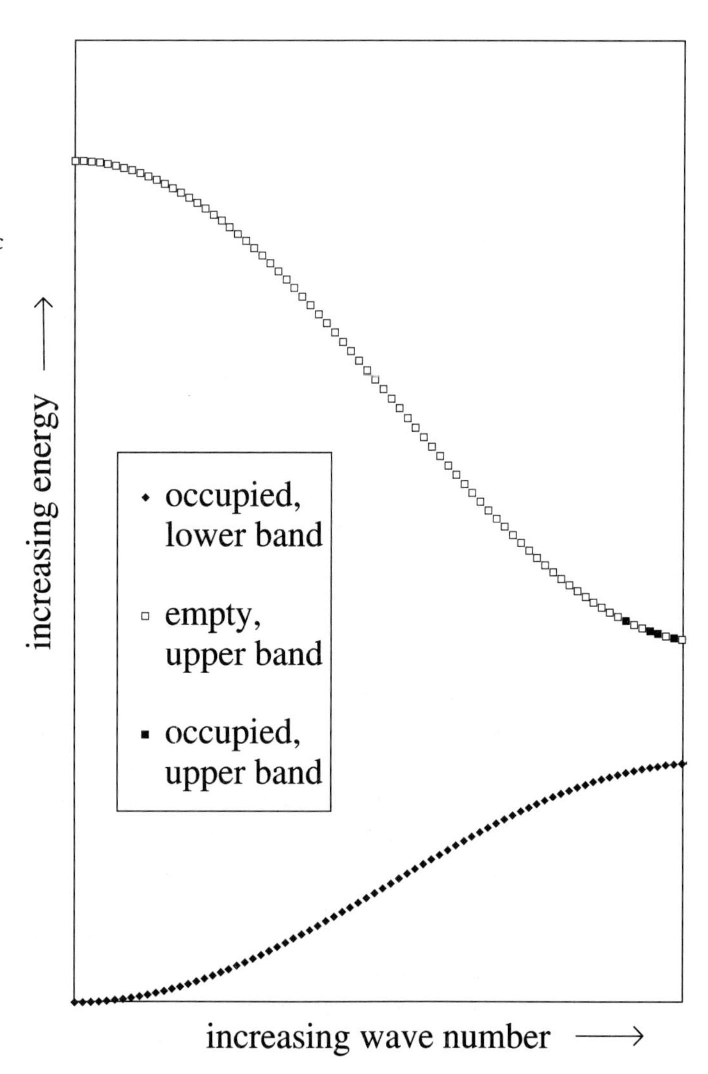

FIGURE VI-17. Energy bands in phosphorus-doped silicon. All states in the lower band are occupied by electrons, just as in pure silicon. The extra electron from each phosphorus atom goes into a state in the upper band, and can contribute to an electric current.

The two kinds of doped semiconductors described above are said respectively to be *n-type* (*n*egative electrons) and *p-type* (*p*ositive holes). Note that such semiconductors can have electrons (or holes) even at the absolute zero of temperature. As the temperature is raised, additional electrons from the lower band gain enough thermal energy to move to states in the upper band, leaving behind empty states in the lower band equivalent to holes, as with the pure semiconductor.

6 Thermal energy of a crystal

We saw in chapter V that changing the temperature of an object changes the energies of motion of the particles that make up the object. The total change for all the particles is the change in the thermal energy of the object produced by the change in temperature. In a metal, the moving particles are the conduction electrons that are reasonably free to move around in the crystal, and the ions that cannot run away from their lattice sites but can vibrate about their average positions. In an insulator (or semiconductor), we have the atoms that again can vibrate about their positions, and we may have some mobile electrons and holes also.

Consider a metal at some temperature T. (I shall use this symbol for the temperature when I want to leave its actual value unspecified; otherwise I shall say that the temperature of something is 300 °C or 800 K or whatever.) The particles in it are the ions and the conduction electrons. The electrons are fermions, and so their distribution among their energy states will be as shown in fig. V-4 (p. 99). If an electron is to take on additional energy as thermal energy because of the temperature, it must have an empty state of energy that is higher by this amount into which to move. The Fermi energy for the electrons in a metal is about a hundred times their average thermal energy. The consequence of this is that only electrons in a very small range of energies (approximately the average thermal energy) around the Fermi energy can take on the thermal energy in addition. Most of the electrons are unaffected by temperature, because it takes more energy for them to move into empty states than can be provided by the thermal energy.

Figure VI-18 shows what this state of affairs looks like in terms of the Fermi surface, which I have taken to be the same as the one in fig. VI-15. Some electrons that were close to the Fermi surface acquire thermal energy and move to states of higher energy just outside the surface. This leaves an equal number of empty states inside the Fermi surface, which behave like holes. You notice that only a thin layer of states at the Fermi surface is affected by temperature. It turns out that our interest is in knowing how much increase in energy takes place with an increase in temperature, rather than the actual energy at any given temperature. This means that we can think of the thermal energy as being used to create electron-hole pairs, an example of which is shown in fig. VI-18 on the right. An electron in state A inside the Fermi surface gains thermal energy and changes to state C outside the surface, leaving an empty state at A. One can divide this gain in energy into two parts: the energy in going from A to B (at the Fermi surface) is

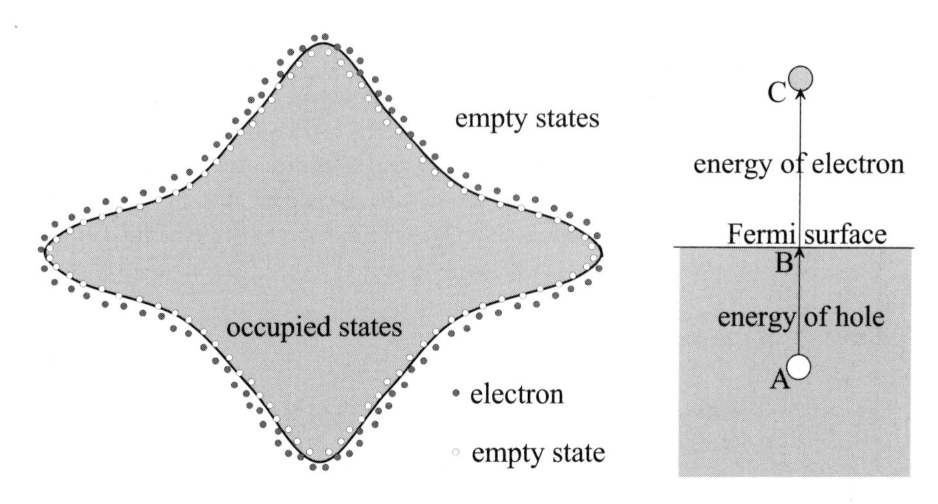

FIGURE VI-18. The same Fermi surface as in fig. VI-15 but now at some temperature above the absolute zero (shown on left), and an electron-hole pair (on right). The thermal energy is very small compared to the Fermi energy at all temperatures of interest. This means that only a few electrons very near the Fermi surface have empty states of slightly higher energy to which they can go after gaining thermal energy (dark circles). These electrons leave behind empty states (open circles), equivalent to holes. The distribution of electrons in all the other states is unaffected by temperature; these electrons have nowhere to go, because all nearby states are already occupied by other electrons. The thermal energy is divided between the electron and the hole as shown on the right and described in the text.

thought of as the energy of the hole, and the energy in going from B to C as the energy of the electron. The thermal energy can create these electron-hole pairs from only a small fraction of the total number of electrons, namely those with energies close to the Fermi energy. This means, as we shall see in chapter VII, that only a small part of the thermal energy of metals is carried by the electrons and holes. The greater part is in the vibrations of the ions.

With insulators (or semiconductors, pure as well as doped), this is even more so, for there is an energy gap to the nearest empty states for the electrons. So even fewer of them make it across the gap with their additional thermal energies, leaving empty states behind. For all practical purposes, the whole thermal energy is carried only by the ions. It is only at very low temperatures, where the total thermal energy is very small, that a significant fraction of it is in the electrons and holes.

7 Phonons

We now look more closely at the thermal vibrations of the atoms (or ions) in the crystal. This is energy of motion, and is therefore connected with the mass of the ion and not its electric charge. We want to find a way of describing this motion quantum-mechanically. We have seen that quantum mechanics is intimately connected with wave motion, and so it is reasonable to see if we can somehow bring waves into the picture of the vibrating atoms in the crystal. If I send a sound wave of a certain frequency down a crystal, and imagine that I could look at the individual atoms, I would find that each atom is vibrating at that frequency. In addition, at any given instant of time the positions of the atoms would trace out a wave of the corresponding wavelength. So there seems to be some promise in trying to connect sound waves with atomic vibrations.

QUESTION How do we know that sound can travel through solids?

ANSWER Because if there is a loud enough noise inside a room, I can hear it from outside even if the doors and windows are shut tight. I can hear the person in the flat upstairs moving furniture about, because the sound comes through my ceiling.

In a solid at some temperature, the atoms are vibrating by themselves even though no sound is fed in from outside. We saw in chapter II that the influence of atoms on one another in a solid is as if they were connected by springs. For simplicity consider a short one-dimensional crystal, with just 11 atoms: very short, indeed. Assume that the atoms at the two ends are clamped, so that they cannot vibrate. Figure VI-19 shows five of the waves that could exist in this lattice, with each wave displaced vertically for clarity. Lines connect the atoms so that the wave shape is easily seen; you can think of the lines as representing the imaginary springs between pairs of atoms. We look first at the lowest wave in the figure, labelled 'fundamental'. There is exactly half a wavelength in this length, and so its wavelength is $2L$ if the length of the crystal is L. The speed v of a wave is equal to its wavelength multiplied by its frequency, and so its frequency f_1 is given by

$$f_1 = \frac{v}{2L}.$$

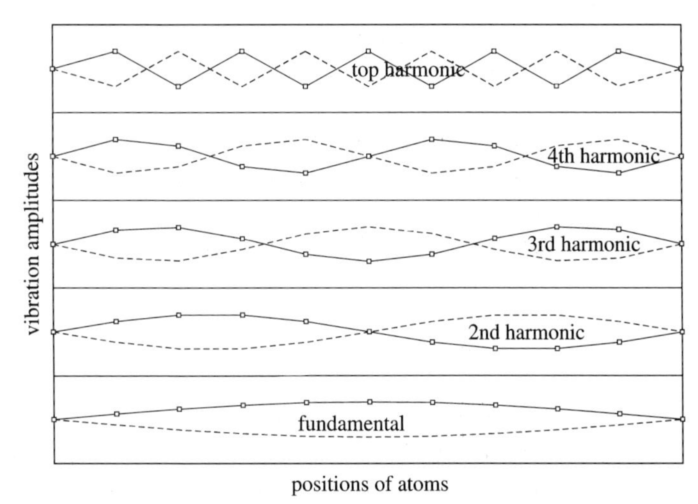

FIGURE VI-19. Some modes of vibration of a one-dimensional crystal with 11 atoms, which are shown as small squares. The lines joining the atoms are to make it easy to see the wave shapes. They may be thought of as indicating the virtual springs connecting the atoms. Five of the possible modes are shown. In each case, the line of atoms is vibrating between the extreme positions shown by the continuous and broken lines.

The next wave, labelled 'second harmonic', has a wavelength equal to L. If the speed of sound is assumed to be the same for all wavelengths, then the frequency f_2 of this wave is

$$f_2 = \frac{v}{L}.$$

or

$$f_2 = 2f_1.$$

A pattern seems to be emerging for the frequencies of the successive harmonics. The third harmonic has a wavelength equal to two-thirds of the length L, and therefore its frequency f_3 is

$$f_3 = v \text{ divided by } \tfrac{2}{3}L \text{ or } \frac{3v}{2L}, \text{ which is the same as } 3f_1,$$

and the fourth harmonic has a frequency f_4 equal to $4f_1$. The harmonics have frequencies given by multiplying the fundamental frequency f_1 successively by the integers 2, 3, ...

There is a simple experiment you can do on a piano that will illustrate this property of the harmonics. The predominant tone you hear when you play the note of middle C has a frequency of 256 vibrations per second: this is the fundamental. The successive octaves of this note have frequencies that are two, three, ... times this frequency. Now gently depress the key an octave above middle C, so that its damper is off the strings but the note does not sound. Then play middle C and let the key up at once. You will hear clearly the tone of the octave, which is the second harmonic of what you just played. This is because when you played middle C, its strings vibrated not only at the fundamental but also at the higher harmonic frequencies. The strings an octave above pick up the sound waves of the second harmonic and begin to vibrate at the same frequency. You get similar results if instead you hold down keys which are two, three, ..., octaves above the note you play. The experiment shows that the tones on the piano contain the fundamental as well as the harmonics.

Now look at the wave labelled 'top harmonic' in fig VI-19. The neighbouring atoms are displaced in opposite directions from where they would be if there were no wave. The wavelength here is just twice the interatomic distance: I cannot have a wave that has an even shorter wavelength. So I conclude that there is a minimum wavelength for the lattice vibrations and a corresponding maximum frequency of vibration. Since the wave number is just the reciprocal of the wavelength, I can equivalently say that there is a maximum wave number that I shall call k_D and a maximum frequency f_D for the waves. I use the subscript D here to recognise Peter Debye, the physicist who first thought of the thermal vibrations of atoms in solids in terms of sound waves.

Each of the vibrations of this lattice has its frequency as well as wave number, and fig. VI-20 shows how the frequency depends upon wave number. I have shown in this figure all the possible modes of vibration, up to the top harmonic. This picture summarises all the information about the vibrational modes of the crystal, except how big the vibrations are, namely their amplitudes, and we consider this aspect next.

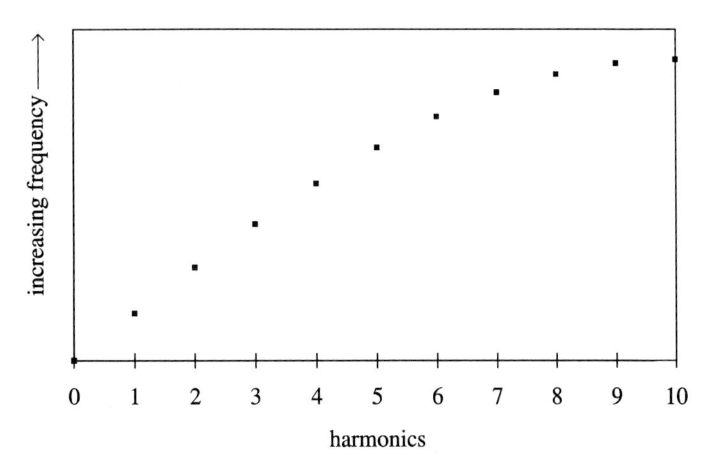

FIGURE VI-20. The frequencies of the different harmonics for the example shown in fig. VI-19. The picture is qualitative, and no scale is shown for the frequencies. As the harmonic number increases (which is the same as saying the wave number increases), the frequency of vibration increases. Since energy is equal to the frequency multiplied by the Planck constant, this picture also shows how energy varies with wave number, and is therefore the dispersion curve for the vibrations of the crystal of fig. VI-19.

QUESTION The speed of a wave is equal to its frequency divided by its wave number. Figure VI-20 shows that this speed is constant for small wave numbers, but becomes somewhat smaller for the largest wave numbers. Why is this so?

ANSWER The smallest wave numbers correspond to the longest wavelengths, and at these wavelengths a single wave extends over many atoms: something like 10^7 or so in a crystal about one cm on the side. So on the scale of the wavelength, the discrete atomic nature of the crystal is smoothed out, and the speeds are the same. For the largest wave numbers, there are very few atoms, much less than ten, in a wavelength. In this case, the discrete atoms must be explicitly taken into account, and the result is that the speed comes out to be less than for very long wavelengths.

Let us look again at one of the modes of vibration, say the fundamental with a frequency f_1. The larger the amplitude of its vibration, the greater is its energy. According to quantum mechanics, this energy cannot take any arbitrary value. It can only take the values $E_1(n)$ given by

$$E_1(n) = nhf_1,$$

where n has one of the values 0, 1, 2, 3,..., and h is the Planck constant. I have neglected here the zero-point energy of the vibrations, because it plays no part in what follows. Thus, as expected, the energy is quantized. This quantum of energy, hf_1, is called a *phonon.* The energy of this mode is determined by the number n of phonons in it: the larger its amplitude, the more energy it has and the greater the number of phonons it has. This applies to each of the possible modes of vibration in the crystal.

So what exactly is a phonon? A phonon is a quantum of energy of the atomic vibrations of the crystal, characterised by a wave vector **k** and corresponding frequency f, and n of these phonons have energy nhf. I can think of the thermal energy of the lattice as if it were contained in a gas of phonons, whose quantum states are specified by

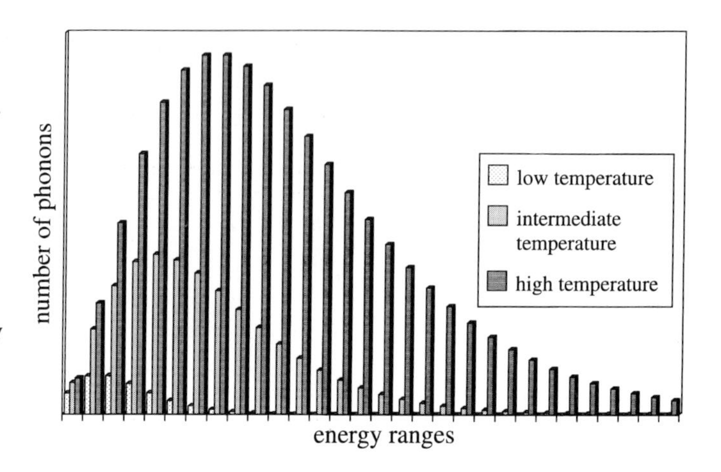

FIGURE VI-21. The numbers of phonons which lie in different energy ranges, at three different temperatures. The picture is qualitative, and no numerical values for the different quantities are shown. The number of phonons is proportional to the height of the bar, and energies increase from left to right. As the temperature rises, the number of phonons in each energy range increases. This means that the total energy of all the phonons also rises with increasing temperature.

k and *f*. Each of these states can be occupied by any number *n* of phonons, the actual number giving the total energy of this state. We recall from chapter V that this is exactly the property of the particles we called bosons, any number of which can be in the same quantum state. We conclude that phonons satisfy the statistics of bosons. The actual number of phonons in each quantum state specified by **k** and *f* is then set by the temperature, and varies as the temperature is varied. I show this variation for three temperatures in fig. VI-21. This is a picture of the phonon distribution, which shows several notable features. The number of phonons, and therefore the energy, in each energy range increases as the temperature increases. At each temperature, there is a particular energy range that has the maximum number of phonons, and the number in higher energy ranges decreases progressively. If you will now go back to fig. V-2 showing the velocity distribution of atoms in a gas, you will see a similarity to this picture, but with one important difference. The total number of atoms in a given quantity of gas does not change with temperature. You can see from fig. VI-21 that the total number of phonons, got by adding up the numbers that are in each of the energy ranges, increases with increasing temperature.

Let me pause now and take stock of what we have done. We began with a crystal whose atoms were vibrating because of the thermal energy of the crystal. These vibrations were equivalent to sound waves, whose energies were quantized as phonons. The phonons belong to the class of particles called bosons. The state of the crystal at some temperature can be described, as far as the atomic vibrations are concerned, by saying that the volume of the crystal is filled with a gas of particles called phonons. There is a kind of symmetry here with the quantum description of electrons confined in a box. In that case, we started with objects that we think of as particles (electrons), and ended by describing them as waves filling the whole box. Here, with atomic vibrations in a crystal, we began with sound waves filling the crystal, and ended up with particles (phonons) moving about in the space occupied by the crystal.

QUESTION Are there any differences between a gas of electrons and a gas of phonons?

ANSWER Yes, many. I tabulate some in table VI-2.

Table VI-2. *Some properties of electrons and phonons*

property	electrons	phonons
mass	yes	no
electric charge	yes	no
statistics	fermion	boson
total number in crystal	fixed by number and kind of atoms	varies with temperature

8 The quantum mechanical crystal

We have come a long way from the crystal lattice of chapter II to the quantum mechanical crystal of this chapter. The only property that we gave the crystal lattice was its translational symmetry, from which emerged other symmetries like those of rotation and reflection. We then put atoms at the lattice points, let the crystal be at some temperature, and used the results of the quantum mechanics of atoms (chapters III and IV) and statistical physics (chapter V) to describe the resulting situation. We find that the atomic nuclei sit at the lattice points. The wave functions of the few outermost electrons are significantly altered from what they were in the isolated atom, and the remaining electrons are nearly unaffected by being in the crystal. It is the outer electrons that contribute to those properties of crystals that are of interest to us. Their energies lie in bands that are separated by energy gaps. Within each band is a characteristic relation between the electron's energy and the wave number of the electron wave. The electrons are fermions, and so a given quantum state can hold at most two electrons with spins up and down respectively. At the absolute zero of temperature the electrons are in successive energy states in the bands, starting with the state of lowest energy. The highest energy band with any electrons at all in it will either have all its states occupied (an insulator), or will have some fraction of the states empty (a metal). As the temperature of the crystal is raised, a few electrons which are in the states of highest energies acquire additional thermal energy, cross the energy gap and go into empty states in the upper band if it is an insulator or semiconductor. If it is a metal, electrons move to nearby empty states in the same energy band. Empty states, equivalent to holes, are left behind in either case. Most of the electrons in both cases do not change their quantum states when the temperature of the crystal is changed.

The effect of temperature on the atoms in the crystal is to set all of them vibrating about their lattice positions. These vibrations are equivalent to the fundamental and harmonics of sound waves filling the crystal. The application of quantum mechanics to these waves leads to the quantization of their energies, and these quanta are called phonons. A detailed study of phonons shows that, just as for electrons, phonon energies also lie in bands with gaps in between. One has a relation between the energy (which is the frequency multiplied by the Planck constant h) of a phonon and its wave number, again analogous to electrons.

An ordinary crystal will thus present one of two appearances, depending on how I look at it. With my eyes aided by a suitable magnifying device, I would see the atoms sitting in a lattice and jiggling about their lattice positions. I have shown a portion of such a crystal in fig. VI-9. I could look at this picture and admire its symmetry, but I would be hard put to conclude anything about its various physical properties – electrical, optical, magnetic, etc.

Now if I visualise the same crystal incorporating the quantum nature of matter, then it will be like the left half of fig. VI-22. There is the volume of the crystal, but inside it are not the atoms, but rather a gas of electrons and holes, mixed up with a gas of phonons, all colliding with one another. I also see curves showing how the energy depends on wave vector for electrons and holes as well as for phonons, as shown in the right half of fig. VI-22. These are the dispersion curves for electrons and holes, and for phonons. Further, if it is a metal, I also see the Fermi surface. It is this view of the crystal that forms the starting point for understanding why it has its particular physical properties.

In going to the quantum-mechanical view of the crystal, I seem to have lost touch with the symmetry of the lattice. Figure VI-22 seems to contain no indication of the translational symmetry of the lattice. It only seems so: you recall that for the phonons in the one-dimensional crystal the largest wave number corresponded to a wavelength equal to two lattice spacings. The detailed working out of the quantum mechanics of the electrons in the lattice leads to a similar limit on the largest possible wave number for the electrons and holes. The essential aspect of translational symmetry is the lattice spacing, which determines how the atoms are spaced out. Thus the existence of such a maximum wave number (or a set of maximum wave vectors for the three-dimensional

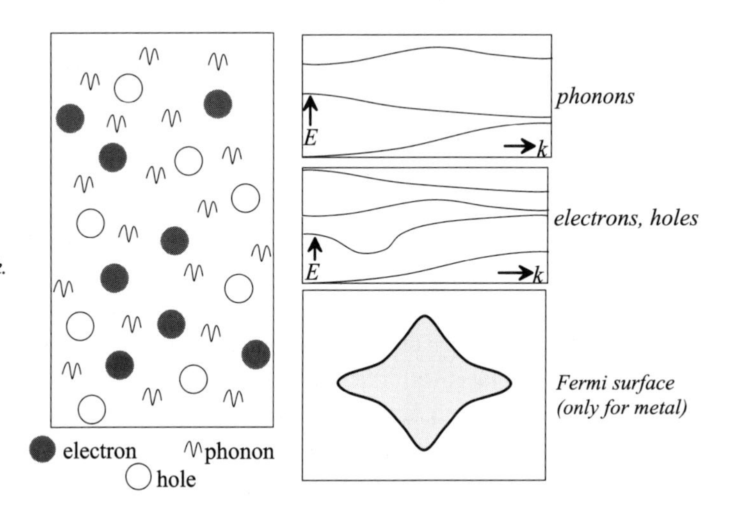

FIGURE VI-22. A crystal of silicon as seen by an eye that gets the quantum mechanical picture. The eye sees a box filled with electrons, holes and phonons moving about. They show wave-particle duality and have dispersion relations for the variation of their energy E with wave number k. If the crystal is a metal, it also has a Fermi surface in addition to these features.

crystal) is related to the translational symmetry of the lattice. In addition, the shape of the Fermi surface for a metal contains implicitly the symmetry of the corresponding crystal lattice. Thus the quantum mechanical view of the crystal not only gives important information about the electrons and about the effect of temperature on the crystal, but also contains in it the crystal's symmetry.

QUESTION In the view of quantum mechanics, how does the crystal of germanium differ from that of silver?

ANSWER Both crystals have electrons and phonons. The difference will be in their dispersion curves, and in the distribution of electrons among their quantum states. This difference explains why their properties are not the same.

9 Summary

Section 1 outlines the task undertaken by us in this chapter: to arrive at a quantum mechanical picture of a crystal in the same sense in which we got such a picture of an isolated atom in chapter IV.

We imagine in section 2 a process of assembling a crystal starting with isolated atoms that are far apart from one another, and gradually bringing them close to one another. When the atoms are far apart, the energy levels for the electrons are the same as for the isolated atom, except that each level is highly degenerate, because it can accommodate one electron from each of the 10^{21} or so atoms that will eventually make up the crystal. In the crystal, on the other hand, this degeneracy is removed, and each atomic energy level becomes a band of closely spaced levels. Each electron now has a slightly different energy, the reason being the influence of electrons and nuclei from neighbouring atoms. There are gaps between adjacent bands, meaning that there are no states for electrons with energy values lying inside these gaps. These effects are most pronounced for the outermost electrons of the atom because they are closest to, and therefore feel most strongly the influence of, neighbouring atoms in the crystal.

The electrons in the crystal occupy these band states beginning at the lowest energy and progressing upwards until all the electrons in the crystal are housed. If at this stage the highest occupied band is exactly full and the next one is separated by an energy gap and empty, we have an insulating crystal. If on the other hand the highest band is only partially occupied, we have a metal. This is explained in section 3, which then goes on to describe some examples of the probability distribution of electrons in actual crystals, which complements the description in terms of energy bands.

The energy states of the electrons have corresponding wave functions, just as with particles in a box and with isolated atoms. We need not worry about how these waves look. Considering that the crystal is a more complex object than the other two, we would expect the wave functions to be more complicated, and this indeed is so. Each wave will still have its wave vector $\mathbf{k}$, and represent a state of energy E. We can make a picture for each band of how E varies with $\mathbf{k}$ from the bottom of the band to the top.

We use such pictures, called dispersion curves, in section 4 to see again how metals and insulators differ from each other. Using the wave vector of the electron and the fact that electrons are fermions, we are led to the picture of a Fermi surface for metals.

In section 5, I introduce the semiconductor as a special form of insulator. We use the pictures of dispersion curves to see how temperature and impurities produce semiconducting behaviour in an insulator. The name itself is suggestive: the ability to carry an electric current for a semiconductor is somewhere between that for a metal (very good) and an insulator (very bad). We find that an otherwise filled band that has a quantum state with no electron in it behaves like a particle, named a hole. Its characteristics like energy, wave vector, charge and spin are the opposite of the missing electron: an anti-electron, so to speak.

I take a first look at what temperature does to the crystal in section 6. Temperature is related to the energy of motion of the electrons, holes and atoms. Raising the temperature increases this energy. The energy of motion is determined by the mass of the moving object, and not its electric charge. In a crystal we have for each atom a few of the outermost electrons that may be wandering around in the crystal. The ions that are left cannot get away from their lattice positions, but can only vibrate with increasing amplitude (which means increasing energy of motion) as the temperature of the crystal is increased. This vibrational energy depends on the mass and not on the electric charge of the vibrating ion. The mass of the ion is practically the same as that of the atom. I can therefore talk about vibrating atoms, even though they may actually be ions. The total of the vibrational energy of all the atoms, plus the additional energy (usually small in comparison to the other) which the electrons and holes pick up due to temperature, is the thermal energy of the crystal.

I give a quantum description in section 7 of the thermal vibrations of the atoms in the crystal. I find that these vibrations are equivalent to sound waves in the crystal, and their energies are quantized following the rules of quantum mechanics. These quanta of energy are called phonons, and they follow the statistics of bosons. They, like electrons, have dispersion curves that show how their energy varies with wave vector.

Finally in section 8, I bring together the different elements of the quantum picture of a crystal that have been developed in this chapter: electrons and holes and phonons, dispersion curves of energy versus wave vector (or wave number), Fermi surface for metals, energy gaps for semiconductors.

We have thus arrived at a quantum-mechanical description of a solid based on its crystalline atomic structure. We are ready now to understand specific properties of solids in terms of this description. We start with a group which gives the word 'solid' its meaning as defined by the Concise Oxford Dictonary: *of stable shape, having some rigidity.* A piece of solid has some shape, and it resists – at least up to a point – attempts to change it. These properties are the easiest to observe experimentally: they need nothing more than the experimenter's eyes and hands, and not the complicated equipment needed with the other properties we consider later. These mechanical properties, as they are called, are the subject of chapter VII.

VII COPPER WIRES AND GLASS RODS

1 Elasticity, plasticity and brittleness

We have looked in the preceding chapters at the atomic structure of solids, described in terms of quantum mechanics. We shall now see how this description explains why solids have the various properties that they exhibit. We begin by looking in this chapter at a group of properties that defines what a solid is as distinguished from a liquid or a gas. The atoms in a solid stay where they are, and do not slide past one another as in a liquid, or fly about in the enclosing volume as in a gas. A solid holds its size and shape, and resists attempts to change them. The ideal solid is a perfect crystal, whose atoms are bound to one another by the electrical forces among their electrons and nuclei. It is as if there were springs connecting each atom to neighbouring atoms. Any stretching or compression produced by an applied force is taken up by the springs, and the solid returns to its original state when the force is removed. The response of a real solid to such a force, however, is more complicated, and I illustrate it with the following experiments.

I take a copper wire and a glass rod, both of the same size. If I grasp the ends of the copper wire and bend it just a little and let go, it springs back to its original shape. If I bend it more than a certain amount and then let go, it stays bent. The same experiment with the glass rod ends differently. The rod will bend a bit initially and recover when I let go, but the attempt to bend it further will cause it to snap into two or more pieces. I find yet another result if I try the experiment with a wire of the kind of steel used to make springs. I can bend this wire a fair amount, but it will spring back to its original shape when I let go. I go back to the permanently bent copper wire, straighten it out and again bend it and so on, repeating the cycle a number of times. I find that it gets harder to change the shape of the wire with each successive cycle: the copper wire begins to resemble more and more the steel spring. Finally, I take the springy copper wire from this last experiment and heat it up to 500 °C, say, and hold it there for some time. This kind of heat treatment is called *annealing*. I test the wire after it has cooled down, and find that it has lost its springiness.

All these materials are made up of atoms, and they are held together by electric forces derived from the charges of the electrons and nuclei. Nevertheless, these experiments show that they respond in different ways to a mechanical force that tries to change their shape. These different types of behaviour can be understood when one looks closely at how the atoms are arranged in real solids, as distinct from the idealised perfect lattices that we have assumed so far. Before I get to that, I summarize the behaviour of a solid when subjected to forces as mentioned above.

The experiments all have something in common: I apply a mechanical force to a sample, and note the resulting change in its shape. For a small enough force, the change is proportional to the force, and is reversible, i.e. the change is always the same for a given strength of force. In this regime the material is said to show *elasticity*. This prevails up to some maximum force, beyond which the deformation of the material does not disappear if the force is taken away. At this point the material has reached its *elastic limit*, and its size and shape will have changed by a few parts in ten thousand. Beyond the elastic limit, the material will either break (*brittleness*), or remain permanently deformed (*plastic deformation*). I can summarize the task now as an attempt to understand the elastic, plastic and brittle behaviour of matter in terms of its atomic structure.

2 The springs

The elastic behaviour of solids is as though there were springs connecting pairs of neighbouring atoms. The electric force between the electrons and the nuclei in a solid is the only mechanism we have to produce the effect of these springs. We saw in chapter VI three different ways that the outer electrons of an atom arranged themselves in solids. They were the following:

1. As free conduction electrons in a metal like silver.
2. An electron leaves each A atom and joins a B atom in an AB compound, thus creating a positive A ion and a negative B ion. The resulting solid is an *ionic crystal*, and sodium chloride (common salt) is an example.
3. Two electrons, one from each atom of a neighbouring pair, come together to bind the atoms together, as in silicon.

I take first an example of a metal, which I assume for simplicity to have one free electron per atom. This electron is a Bloch wave which is spread throughout the whole solid. However, the solid is electrically neutral, and therefore on the average there is one electron in the space around each ion. I show this space bounded by broken lines around four of the ions in the lattice in fig. VII-1, part (*a*). The polygons formed by the broken lines are all identical, because of the translational symmetry of the lattice. We therefore need to find the energy of just one ion and its associated electron; the same result will hold for all other atoms in the crystal. The ion and electron form a kind of atom, but the electron is free to move from one ion to the next. The physicists Eugene Wigner and Frederick Seitz worked out the quantum mechanics of such an atom. They found that the electron stayed closer to the nucleus than it did in the isolated atom, and therefore had a lower energy. This means that the metal is stable, and does not disintegrate into a pile of free atoms. The difference between this energy and the higher energy of the isolated atom has to be supplied from outside if one is to tear an atom away from the solid.

QUESTION Why does the electron have a lower energy when it stays closer to the nucleus?

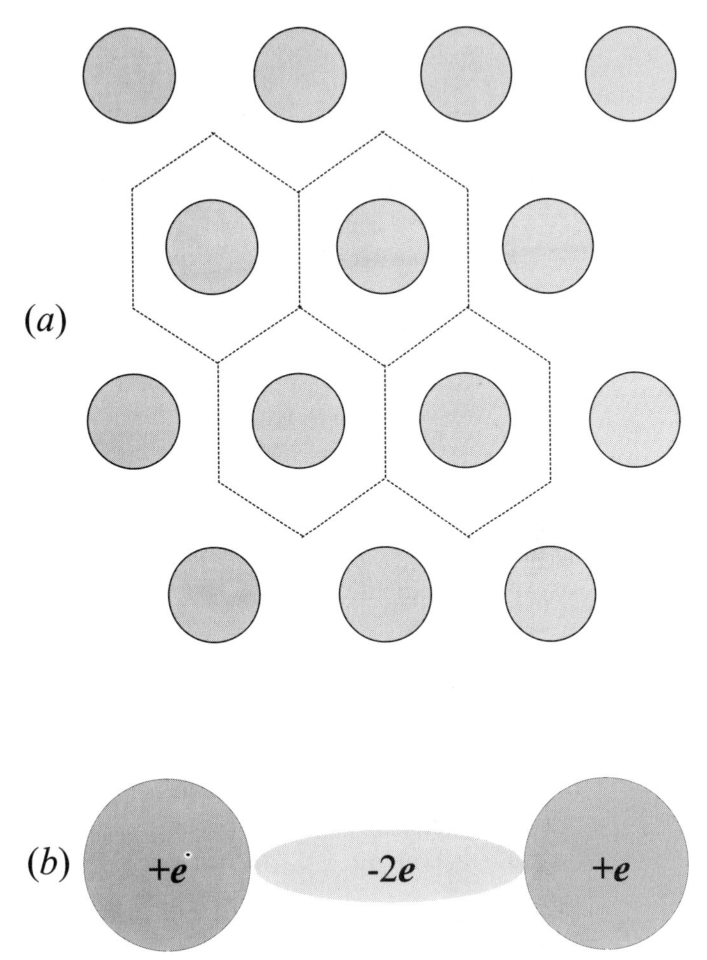

FIGURE VII-1. Part (*a*): Two-dimensional lattice of atoms in a metal with one free electron per atom. An identical space around each atom, like the four that are shown, is imagined to be enclosed by broken lines. One free electron is to be found in each of these spaces, which are called Wigner-Seitz cells. Part (*b*): The bond between two neighbouring silicon atoms in the crystal. An electron leaves each atom and is mostly to be found in the space between the two ions which are left behind.

ANSWER The potential energy of two charges is equal to their product divided by their distance apart (cf. p. 65). Because the electron and the nucleus are oppositely charged, the energy is negative, and becomes more negative, i.e. lower, when the electron is closer to the nucleus.

In an ionic crystal like sodium chloride, the electric force of attraction between the positive chlorine ions and the negative sodium ions holds the crystal together. This effect is illustrated in figs. VI-7 and VI-8. Common salt dissolves easily in water because of the electrical properties of the latter. They cause the attraction between the sodium and chlorine ions to decrease by a factor of 80, and make it too weak to hold them together.

In silicon, whose crystal structure is shown in fig. VI-9, there is a bond between each pair of ions consisting of one electron from each atom. This is illustrated in fig. VII-1, part (*b*). The charge distribution of $+e$ on each of the ions and $-2e$ in the electron cloud between the ions causes electric forces which are as if a spring connected the two ions. If the normal interatomic distance is changed by applying a force, the charge distribution changes in such a way as to oppose this force.

The three kinds of bonding between atoms in solids which I have described have these things in common: each is based on the quantum mechanical description of the atom, is derived from the force between electric charges, and involves a balance between attractive and repulsive forces between neighbouring atoms when they are at the observed lattice spacing apart.

QUESTION Can you clarify the third point above?

ANSWER Since there are both positive and negative charges in the atoms, there must be both attractive and repulsive forces at play. Each atom is sitting still in the crystal, apart from its zero-point and thermal vibrations. There is therefore no net force on it. This means that the attractive and repulsive forces must cancel each other out. Stretching the crystal changes the distribution of charges in such a way that the attractive force dominates, whereas compressing the crystal causes the repulsive force to be the larger.

3 Dislocations

I start with a perfect crystal of some element. I take it to be two-dimensional, so that I can draw pictures that are easier to follow than if it were three-dimensional. I show in part (*a*) of fig. VII-2 such a crystal. The springs between neighbouring atoms in the

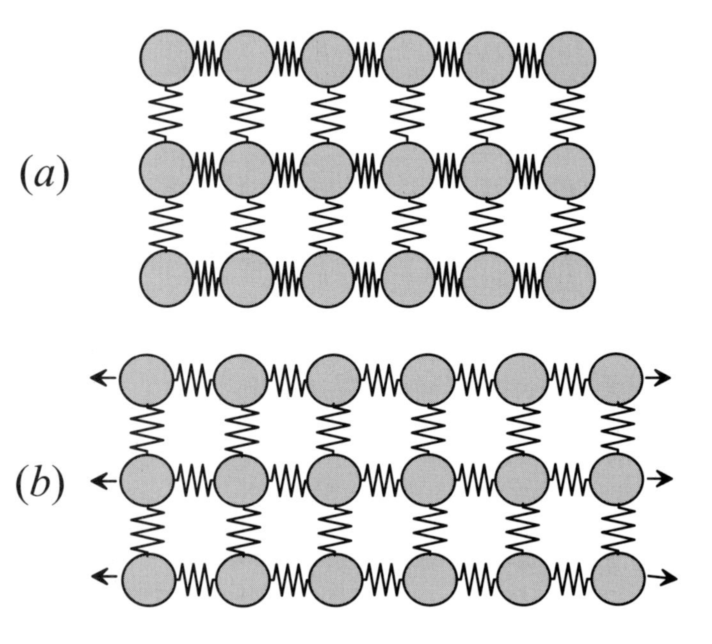

FIGURE VII-2. A crystal is depicted by rows of atoms interconnected by springs which stand for the forces which hold the atoms together. Part (*a*) represents the crystal with no force applied. In part (*b*), stretching forces (shown by the two sets of arrows) are applied to the ends of the crystal, causing the springs in the direction of the forces to stretch and those at right angles to shrink, producing a corresponding deformation of the crystal. When the forces are removed, the crystal returns to the state shown in part (*a*) if the elongation was within the elastic limit.

horizontal and vertical directions look different, to suggest that the crystal is anisotropic. In part (*b*) the crystal is stretched in the horizontal direction as indicated by the arrows. The horizontal springs are then elongated, the other springs are compressed, and the crystal is correspondingly deformed. When the force is removed, the springs as well as the crystal return to the state shown in part (*a*). This is the *elastic* response of the crystal. The same stretching force acting in the vertical direction would produce a different elongation of the crystal: the crystal shows anisotropy in its elastic properties, because the springs in the two directions are different.

We now come to the plastic region, where a strong enough force produces a permanent deformation of the solid. The underlying rearrangement of atoms is pretty complicated, and so I consider a very simple deformation, illustrated in fig. VII-3. I

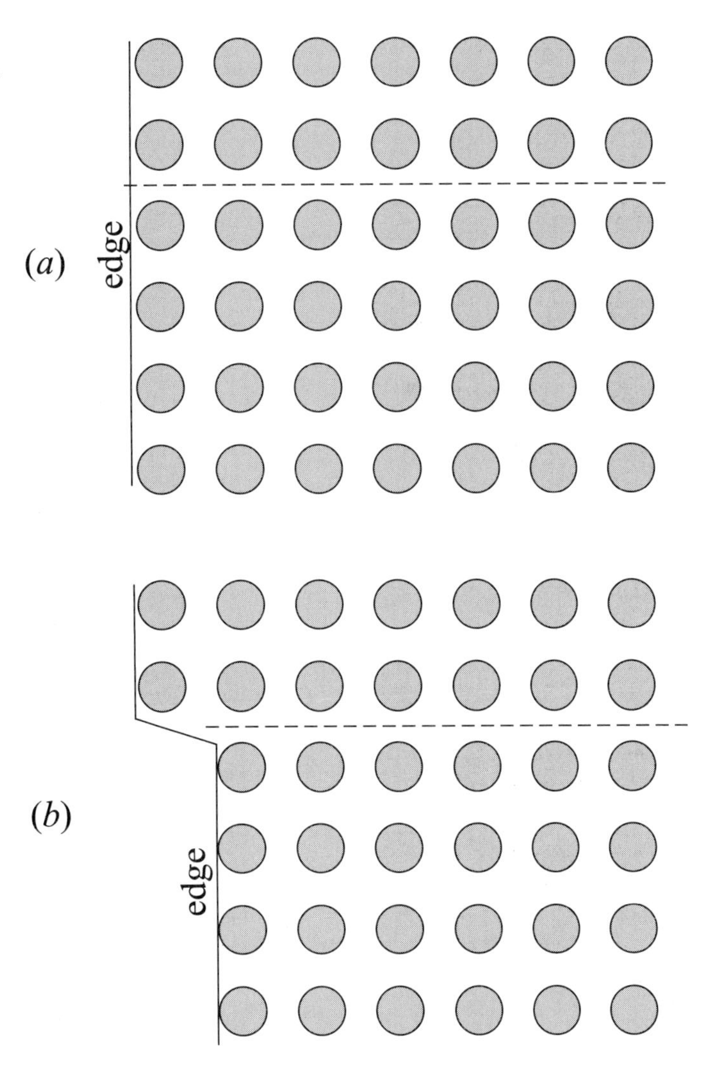

FIGURE VII-3. Atoms near the edge (shown as a vertical line on the left) of a two-dimensional crystal. The part labelled (*a*) is before, and (*b*) is after, a small deformation. To go from (*a*) to (*b*), all the atoms below the horizontal broken line must be moved as a unit to the right by one lattice spacing.

show in part (*a*) the atoms in a small portion of a perfect two-dimensional crystal near its edge. Now suppose I want to deform it, so that it ends up looking like part (*b*) of the figure. One can think of a large deformation as made up of a large number of such small deformations. To produce one such deformation, I would have to move all the atoms below the horizontal broken line simultaneously one lattice spacing to the right. This would require the simultaneous breaking of all the bonds between atoms on either side of the broken line in part (*a*), and then reuniting the two pieces as in part (*b*). The initial breaking of bonds requires a very large amount of energy.

QUESTION How do I know that this operation would require a large amount of energy?

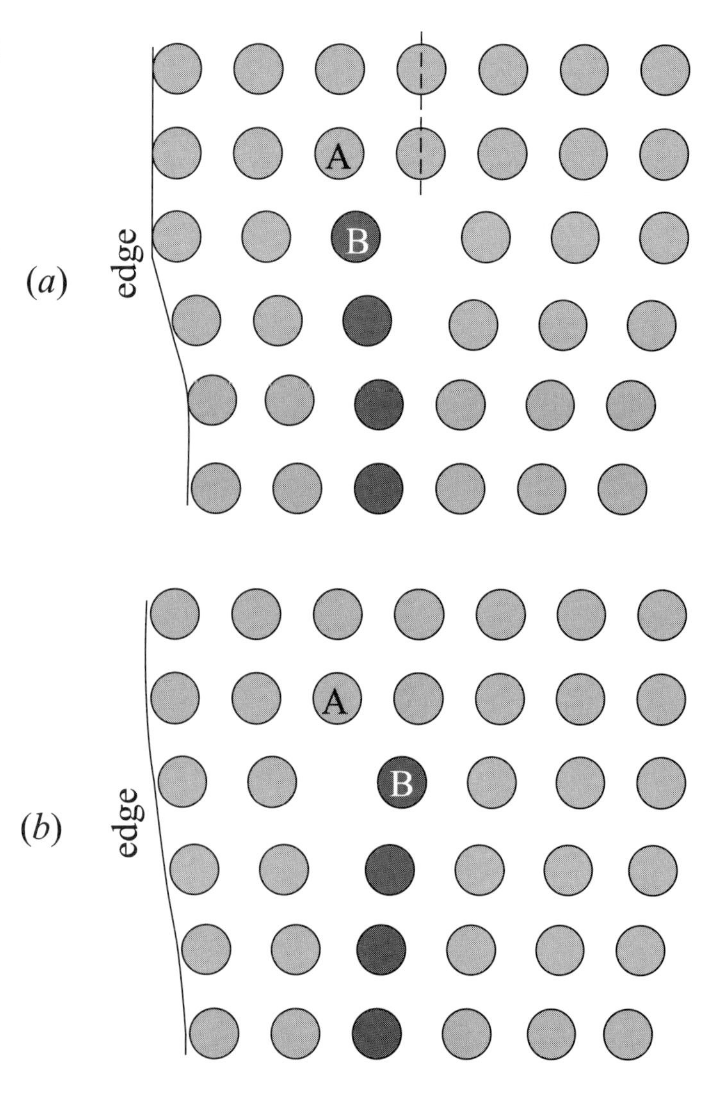

FIGURE VII-4. Part (*a*) shows a crystal with a dislocation in the centre. The atoms in its immediate neighbourhood are only slightly displaced from where they would be in the perfect crystal, and the atoms further away are practically unaffected. The broken line marks the line of atoms which ends at the dislocation. In part (*b*), the atoms marked darker grey have moved slightly to the right, and the dislocation has moved to the left.

ANSWER Think of trying to pull apart a piece of solid into two pieces; this also requires a similar breaking of the bonds between all the atoms on either side of the fracture.

So a perfect crystal cannot be easily deformed in the manner shown above. However, real crystals are not perfect crystals. As they are being formed, not every atom settles down in its legal lattice site. Mistakes are bound to occur, and I show one such in part (*a*) of fig. VII-4. There is a small empty region in the middle, whose volume is about what would normally be occupied by one atom, surrounded by atoms that are only slightly displaced from the exact lattice positions. A very small force, enough to break just one bond, the one between the atoms marked A and B, will cause the row of

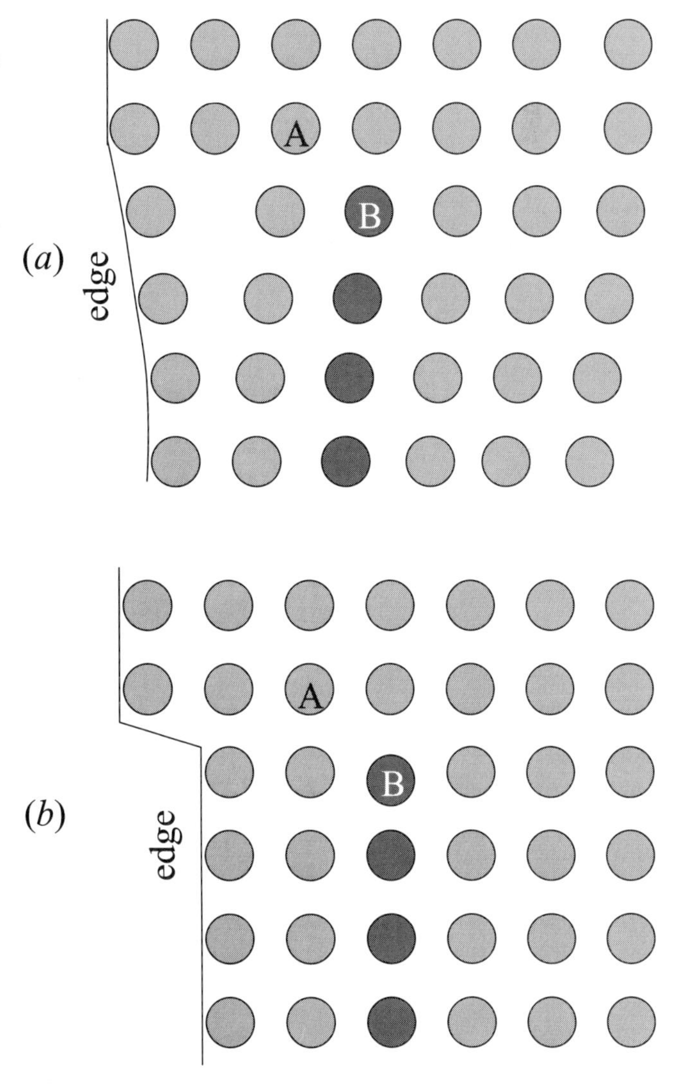

FIGURE VII-5. The same crystal as in fig. VII-3, but now the dislocation has moved one more step to the left in part (*a*), and finally disappeared at the surface in part (*b*), leaving behind a deformed crystal. The same atoms are marked darker grey as are in fig. VII-4.

atoms marked darker grey to flip to the right and produce the situation shown in part (*b*) of the figure. If the force continues, the process repeats itself, as shown in part (*a*) of fig. VII-5. In the end one gets the deformed crystal as in part (*b*) of the figure.

What I have called a mistake in the crystal is called by physicists a *dislocation.* I have described the process as movements of the atoms. If you look at the figures again, you can see that I could equally well have spoken of the movement of the dislocation to the edge, where it vanishes. Such a movement of a single dislocation thus produces a small but permanent deformation of the crystal. One can imagine that the movement of millions of millions of dislocations would produce the plastic deformation that the bent copper wire underwent. Many dislocations are formed during the freezing of the liquid, and more are created during the deforming of the solid.

Figure VII-6 shows a view of the surface of a single crystal of platinum metal, made with an instrument called a scanning tunnelling microscope. It is capable of producing enough magnification so that single atoms appear as circular blobs. There are two dislocations at about the same point in the lattice. Straight lines mark the two lines of atoms which end at the dislocations (compare with part (*a*) fig. VII-4).

When the copper wire is repeatedly bent and straightened, we find that it offers increasing resistance to further deformation. More and yet more dislocations are created, and move, during the bending and straightening. They form an increasingly tangled network that finally inhibits the further movement of the dislocations. We say that the copper wire has undergone *work-hardening.* If now the wire is annealed, i.e. heated

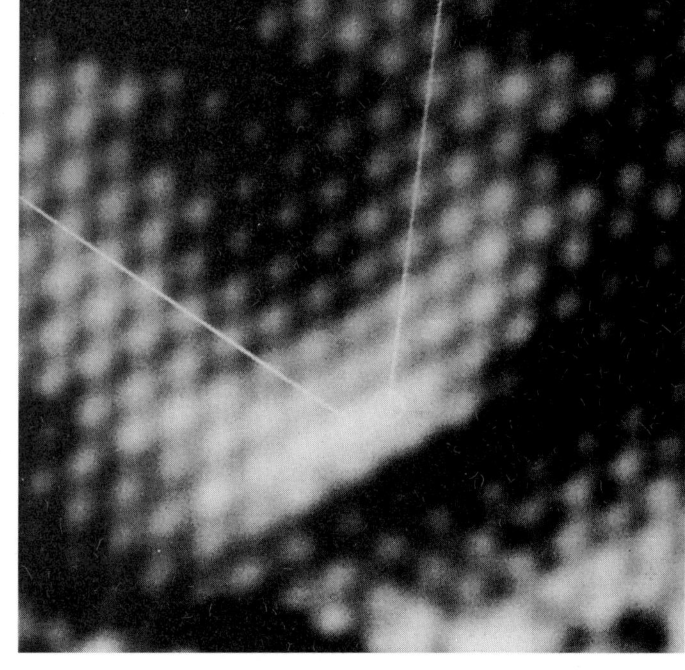

FIGURE VII-6. Photograph of the surface of a single crystal of platinum metal, made with a scanning tunnelling microscope. Each circular blob is an atom, whose diameter is about 3 angstroms. There are two dislocations in the picture. Each of the two straight lines ending in circles marks a line of atoms that ends at a dislocation. Note the similarity to fig. VII-4, part (*a*). (Courtesy of Dr. Michael Hohage, Forschungszentrum Jülich, Germany.)

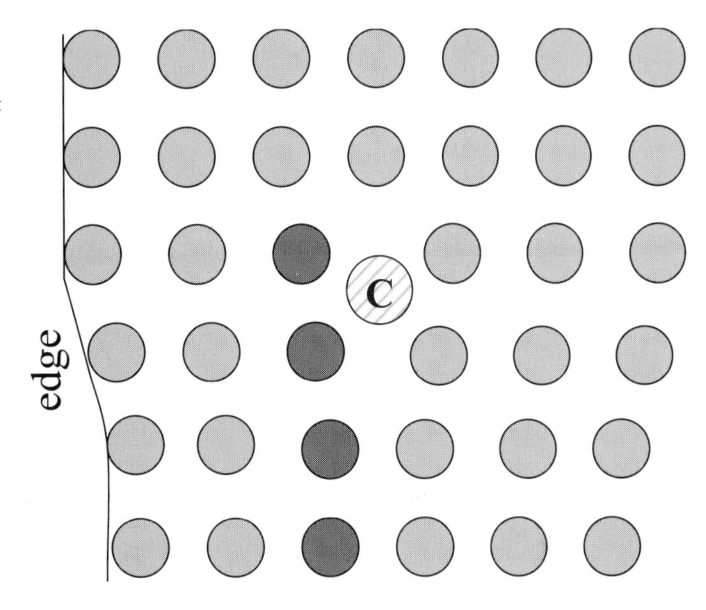

FIGURE VII-7. An alloy atom C which has trapped a dislocation. A movement of the dislocation like that in fig. VII-3 now requires a movement of the atom C also, and this needs considerably more force than when

to a high enough temperature, the increased amplitude of thermal vibrations of the atoms makes it possible for dislocations to break loose from the tangled network and move about. Those that reach the surface of the wire will disappear. The net result is that the work-hardening disappears, and the wire can again be bent easily.

A pure metal can be hardened by another method also, namely by alloying. Pure copper is soft, but an alloy of copper with some zinc is hard enough to be made into pots and pans: it is called brass. Similarly, pure iron deforms too easily to make it suitable as a construction material, but alloying it with a small amount of carbon produces a kind of steel that is widely used in the construction of buildings, bridges and so on. The basic reason for such hardening is that the atoms of the alloying element along with some of the atoms of the host element take up such positions that they act as traps for dislocations and inhibit them from moving, as illustrated in fig. VII-7. This effect of alloying is known as *solution hardening* or *precipitate hardening*, depending on whether the alloying atoms are distributed uniformly or lie in small clusters. By a judicious combination of the hardening methods, one can produce for example the steels that stand large amounts of elastic deformation and are used to make springs.

4 Fracture

There are materials like glass, which deform elastically a little bit, but suffer fracture upon further deformation. We know that the atoms in glass are not in regularly ordered positions as in a crystal. It is therefore not possible to imagine in glass a dislocation such as is illustrated for a crystal in fig. VII-4. It would seem that the only way for glass to deform plastically is for large numbers of interatomic bonds all to be broken at the same time and then reformed soon after: a highly unlikely process that also requires a large amount of energy. How then does a glass rod fracture when bent beyond a certain point?

I start by reminding you of an experience which most of you have had: trying to tear open the plastic sheet in which things like audio-cassettes, new books, nuts, and innumerable other things are packed these days. One cannot tear the plastic sheet at any arbitrary point along its smooth edge. Rather, one must first look for the small nick in the edge that has been helpfully put there by the manufacturer, and start there. I illustrate the process in fig. VII-8 and explain it in the caption.

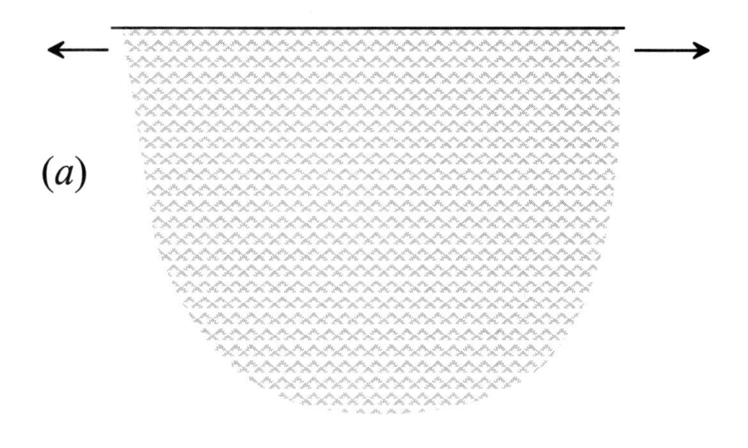

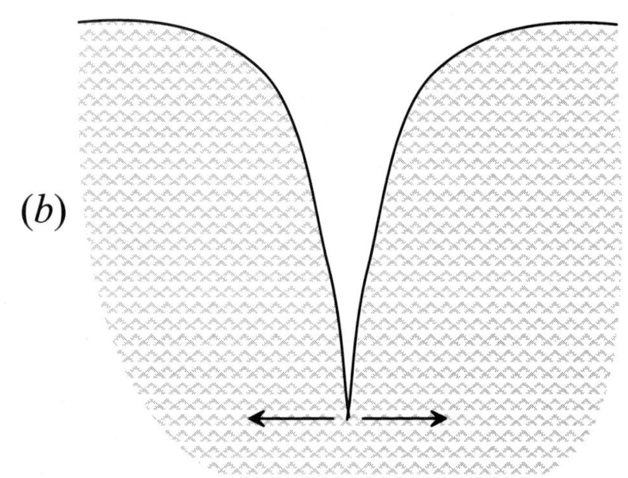

FIGURE VII-8. The tearing of a plastic sheet. The magnification is supposed to be such that the molecules show up as wiggles on the sheet. (*a*) The force, applied to the edge of the sheet at the places shown by the arrows, is divided among many molecular bonds in parallel, and the resulting force on each bond is too small to break it. (*b*) If the edge of the sheet has a sharp notch, the force is concentrated at the tip of the notch and acts on only a few bonds there. The force on each bond there can then become large enough to break it, causing the notch to become deeper, and the process continues.

A similar mechanism is at work when a glass rod is fractured by bending it. Although its surface may look smooth to the eye, there are always present some microscopically small notches on the surface. When the rod is bent, the concentration of force at one of these notches reaches a level at which it begins to grow, and the rod fractures there.

QUESTION How can I get a glass rod to break at a particular point along its length by bending it?

ANSWER By making a scratch at that point with a glass-cutting knife, holding the rod on both sides of the scratch and then bending the rod. The force is then mostly concentrated at this scratch, and the rod will break there. This method is used by glassblowers to break glass rods and tubes to a desired length.

5 Summary

A given solid responds to a mechanical force applied to it in different ways: it deforms slightly and reversibly, or permanently, or it breaks. I describe these different behaviours, called *elastic, plastic*, or *brittle*, in section 1.

In section 2 I show how the springy binding between atoms occurs in metals like silver, ionic crystals like salt, and semiconductors like silicon.

I introduce in section 3 the idea of a *dislocation*, which is a small deviation from the perfect arrangement of the atoms in a crystal. Dislocations are formed naturally during the freezing of the solid from the melt, and are also created when the solid is being deformed. I show how the motion of dislocations leads to plastic deformation. Such motion is inhibited by alloying: the alloying atoms trap the dislocations and prevent them from moving.

I consider the brittle behaviour of glass in section 4. It is caused by the presence of very small scratches on the surface. The applied force becomes concentrated at these scratches, and finally becomes strong enough at one of them to break the molecular bonds locally, leading to brittle fracture.

I have used pictures of a two-dimensional solid to illustrate several points in this chapter. The same pictures, suitably modified, hold for three-dimensional solids too. For example, a dislocation in two dimensions causes a departure from perfect order over an area surrounding one atomic site. A dislocation in a three-dimensional solid causes a similar departure over a volume surrounding a line of atomic sites.

Dislocations are a departure from perfect atomic order in a crystal, and are essential for an understanding of the mechanical properties of solids. However, the vast majority of atoms even in a deformed piece of matter are still in their ordered crystalline positions. Mechanical deformation involves bodily movement of atoms from one position to another, but there is no such movement in the case of the other properties that we shall be considering. These properties arise from two features: the perfect crystalline arrangement of the atoms and the resulting electronic band structures and phonons. The dislocations cause only a small perturbation of these features, and do not affect the essential aspects of such properties, like the one that we consider next.

Temperature is a very pervasive thing in everyday life. The sense of something being warm or cold is acquired very early in life, certainly before understanding something about electric charge, say. So it seems reasonable to take up some thermal properties first, before going on to things electrical.

VIII SILVER SPOONS AND PLASTIC SPOONS

1 Some preliminaries

When I stir hot coffee with a silver spoon, I find that the handle becomes warm very quickly. If on the other hand I use a plastic spoon, its handle stays cool. I conclude that silver conducts heat better than plastic. I find from experience that I can divide materials into two classes: those that conduct heat well, and others that do not. We know that metals like silver, copper or aluminium are good conductors of heat and also of electric currents, while materials like glass or plastics are poor conductors of both. So we may expect that there is something common to the ways in which materials conduct electricity and heat, and we shall see later that this is the case.

In order to understand clearly what we mean by the conduction of heat, consider a simple example illustrated in fig. VIII-1. A bar of some material, silver or glass for example, connects two chambers of which one is hot and the other is cold. If I measure the temperature of the bar along its length, I find that it is hot at one end and becomes gradually cooler as I go along its length. What I have here is an idealised version of the spoon in the cup of coffee. Energy in the form of heat is being continually conducted from the hot chamber along the bar to the cold chamber.

QUESTION How do we know that energy is actually moving in this manner?

ANSWER If I were to replace the cold chamber with a block of ice, I would find that the ice gradually melts because of the thermal energy it is receiving from the bar.

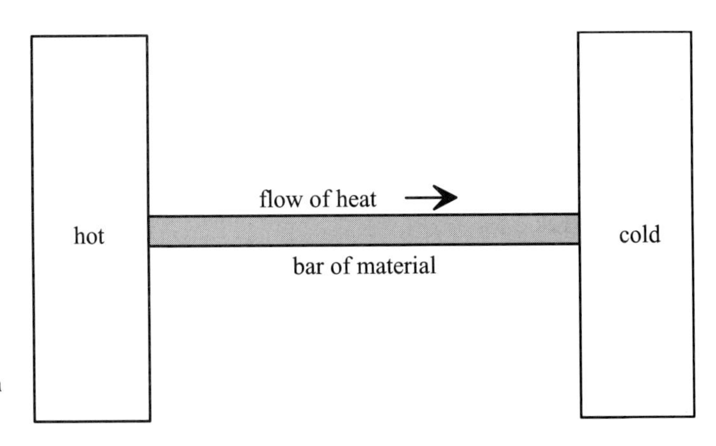

FIGURE VIII-1. Thermal energy is conducted in the bar of material from the hot end to the cold end.

I now do another experiment, where I connect two blocks of ice with the bar after first cooling it to the temperature of ice. I conclude in this case that no heat flows from one to the other through the bar, because I see that the blocks do not melt. I have assumed in all this that no heat flows between the outside and the blocks or the bar.

These two experiments lead to an important conclusion: *Thermal energy flows along a bar only when its two ends are at different temperatures.* It is the difference of temperature which drives the flow of heat along the bar.

Now I do a third experiment. I take a metal spoon, and while holding it at one end between my fingers, I place the other end in a cup of hot tea. That end of the spoon immediately gets hot, and the heat travels slowly down the spoon, as I can tell by touching it at various points at successive times. It takes seconds, perhaps even minutes, for the changes of temperature down the spoon to become evident. I can make a pic-

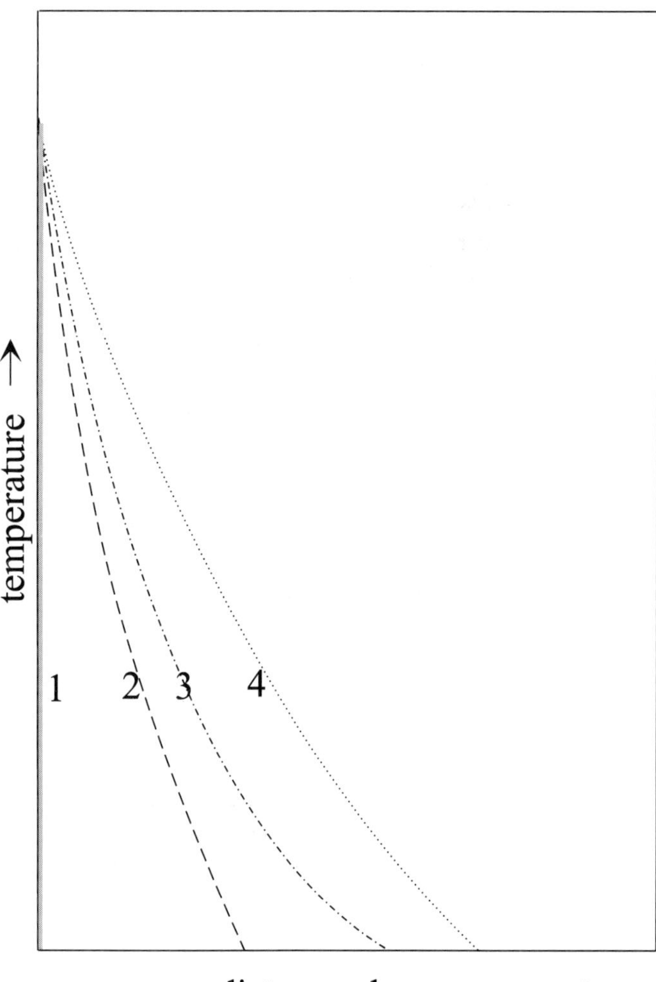

FIGURE VIII-2. The temperature along the length of a metal spoon at four successive instants of time, labelled 1, 2, 3, and 4 respectively, after one end has been dipped in hot tea. Curve 1 shows the situation at that instant: that end is hot, but the rest of the spoon has not yet received the message, so to speak. Curves 2, 3, and 4, at successively later times, show how heat travels slowly down the spoon.

ture, as in fig. VIII-2, of how the temperature along the spoon varies at different instants of time. The curves labelled 1 to 4 show the temperatures at the initial moment of heating and at three later successive times respectively. The sharp rise of temperature which occurred at one end travels slowly down the spoon.

I summarize what we have learnt from these three experiments about the flow of heat down a bar of material: a difference of temperature is necessary in order to produce heat flow, the flow is from the hot end to the cool end, and is rather slow. We would like now to explain these features using the picture of a solid that we developed in chapter VI. To do so, we need two new ideas: a measure of how much thermal energy must be fed into the material in order to change its temperature, and a way of taking account of the collisions which the electrons and phonons suffer as they go whizzing about in the solid. I explain these ideas in the next two sections.

2 Specific heat

When I supply a certain amount of energy in the form of heat to a piece of material, it becomes hotter. If I repeat the experiment with different pieces, all the same mass but made of different materials, then I find that the rise in temperature is different for each material, even though the amount of thermal energy was the same. To put it conversely, it takes different amounts of thermal energy to produce the same rise in temperature in different materials. I illustrate this in table VIII-1 showing the amounts of thermal energy (in joules) needed to heat one gram of different materials from 20 °C to 21 °C. This quantity is called the *specific heat per gram* of the material. Since we know from other experiments the number of atoms in one gram of each material, we can also calculate the specific heat per atom, and this is shown in the second column of the table. The table shows that the specific heat per gram is different for different materials, whereas the specific heat per atom has about the same value for the first three, and is lower for diamond. I shall return to this point presently, but first we look at what happens to the electrons and phonons in a solid when some thermal energy is added to it.

Table VIII-1. *Specific heats at room temperature of some materials*

material	specific heat per gram (joule per °C)	specific heat per atom (10^{-23} joule per °C)
copper	0.39	4.1
silver	0.24	4.3
lead	0.13	4.48
diamond	0.51	1.05
water	4.2	12.6 (per H_2O molecule)

I start with the thermal energy that is in the phonons, representing the vibrations of the ions (or atoms) in the crystal. I assume that the crystal is at 20 °C (which is the same as 293 kelvin). I refer to this as room temperature, and denote it by the symbol T. You recall that there is a maximum phonon frequency f_D (p. 126). If this frequency is such that the energy $k_B T$ is much larger than the phonon energy hf_D, then we find from the statistical physics of phonons that the average thermal energy per atom is just $3k_B T$. The symbol k_B is the Boltzmann constant, a constant of nature with the value 1.38×10^{-23} joule per degree. At a temperature which is one degree higher, the thermal energy per atom is $3k_B(T + 1)$. So the extra energy per atom needed to raise the temperature by one degree, which we have called the specific heat per atom, is the difference between these two energies, or just $3k_B$, which is 4.14×10^{-23} joule per degree. Thus the specific heat per atom should have this value for all materials. The table above shows that this is approximately true for copper, silver and lead, but not for diamond. The reason for this deviation for diamond is that its f_D is much larger than the f_D for the other three materials, and one must go to much higher temperatures to get $k_B T$ to be greater than hf_D. When this condition is not met, the statistical physics of phonons tells us that the specific heat is smaller, and in fact heads towards zero as the temperature approaches the absolute zero. All these predictions are borne out by experiments.

Looking at the table again, we see that the specific heat per atom for the three metals is in fact slightly larger than the value of $3k_B$ which comes from the phonons. This excess is partly due to the presence of conduction electrons in the metals, which can receive thermal energy and move to neighbouring quantum states that were previously unoccupied, and thus add to the specific heat. We now estimate this electronic contribution.

QUESTION The electronic contribution to the specific heat is absent or at best negligible in insulators. Why?

ANSWER The electrons in insulators have no unoccupied quantum states near enough in energy to which they could move by absorbing thermal energy.

Let us consider a metal that has N atoms per cc. Their thermal vibrations, as we saw, give a phonon specific heat which I shall call C_L:

$$C_L = 3Nk_B.$$

The subscript L reminds me that this comes from the *l*attice. Now suppose that each atom sets free one electron; these electrons then form a Fermi surface. At 0 K, all quantum states are occupied by electrons up to a maximum energy called the Fermi energy E_F, which is about a thousand times the thermal energy per atom at room temperature. At a temperature T, the additional energy which an electron can take on as thermal energy is about $k_B T$, which is a very small fraction of the Fermi energy. As shown in fig. VI-18 (p. 124), this means that only those electrons whose energies are less than the Fermi energy by about $k_B T$ can move to empty states of higher energy leaving empty states, equivalent to holes, behind. The number of such electrons and holes is therefore a fraction (k_BT/E_F) of the total number N. Each of these has on the average an energy k_BT, and so their total thermal energy is equal to their number, $N(k_BT/E_F)$, multiplied by k_BT, or

$$Nk_B^2T^2/E_F.$$

If now I warm the crystal by one degree from T to $T+1$, the energy of the electrons and holes increases to

$$Nk_B^2(T+1)^2/E_F.$$

The difference between these two energies is the energy per cc that must be fed to the electrons and holes in the crystal in order to raise the temperature by one degree, and is therefore the electronic specific heat, which I denote by the symbol C_E (the subscript E because this comes from the electrons and the associated holes):

$$C_E = Nk_B^2(T+1)^2/E_F \textit{ minus } Nk_B^2T^2/E_F,$$

which is very nearly the same as

$$C_E = 2Nk_B^2T/E_F.$$

I have used a little bit of algebra to get to the last step, and it is easy to verify it by putting in some numbers. We took T to be 293 degrees and so we must subtract 293 multiplied by itself from 294 multiplied by itself; the result is 587, which is very close to twice 293.

So we now know how much energy must be given to the phonons, and how much to the electrons and holes, in order to raise the temperature of the crystal by one degree. The specific heat of the crystal as a whole is got by adding these two. We can find out their relative magnitudes by forming the fraction C_E/C_L:

$$C_E/C_L = 2Nk_B^2T/E_F \text{ divided by } 3Nk_B$$

which is equal to $2k_BT$ divided by $3E_F$. Since the Fermi energy is a thousand times the thermal energy, this means that the electronic specific heat is about a thousandth of that due to the phonons. We conclude from this that at room temperature the electronic specific heat is a very small fraction of the total specific heat. Most of the energy needed to warm the crystal goes to the phonons, and only a very small part goes to the conduction electrons, in the case of a metal. In an insulator or semiconductor, there are no free electrons (or they are at best negligible in number), and practically all the energy goes to the phonons.

3 Mean free path

I go now back to the bar of material illustrated in fig. VIII-1. Energy in the form of heat is flowing steadily from the hot end to the cold. I want to describe this in terms of the electrons, holes and phonons. They are the entities which contain the thermal energy

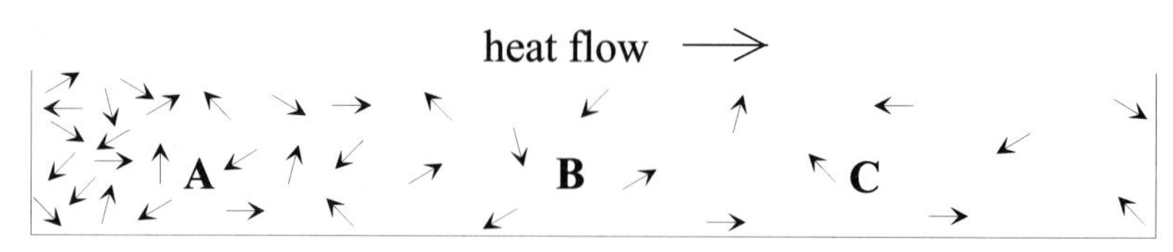

FIGURE VIII-3. Phonons in a bar along which heat is flowing. Although phonons of all frequencies are present, only those of a single frequency are shown. Each arrow represents a phonon travelling in the direction of the arrow. The density of phonons, i.e. the number in a given volume, decreases gradually from the hot end to the cold end. This is what eventually leads to heat flow from hot to cold.

and therefore they must be responsible for its transport down the bar.

Consider a bar of insulator first, in which there are only phonons to carry the thermal energy, for there are no free electrons in it. We shall see later that what we find out about the phonons will apply to the electrons and holes in a metal also, with suitable modifications. Figure VIII-3 shows the bar. The arrows represent phonons of one of the many frequencies present, and remind us that they are speeding about in all directions with the velocity of sound in the material. The density, i.e. the number of phonons per unit volume, gradually decreases as one proceeds from the hot end to the cold end. Exactly the same holds for phonons of all frequencies in the bar, and so the results we get for the phonons of a single frequency will apply to all phonons.

QUESTION Why is the density of phonons higher at higher temperatures?

ANSWER The higher the temperature, the more thermal energy there is in a given section of the bar. This energy is contained in the phonons, and so there must be more phonons in this section. You can see this in fig. VI-21 (p. 128).

Consider a phonon towards the hot end of the bar in the region marked A. It will travel a certain distance before it collides with another phonon. Suppose that this collision takes place in the region marked B. The collision changes the phonons to other phonons whose total energy is equal to that of the colliding phonons because the total energy is conserved. Now there is more energy, and so there are more phonons, in the region of the collision than is prescribed by the temperature there. These surplus phonons continue on until the same thing happens to them further down the bar. There is at the same time a flow of thermal energy in the opposite direction at B, due to phonons arriving there from a region like C which lies towards the cooler end of the bar. The net flow of energy is then the difference between these two.

The distance travelled by a phonon before it suffers a collision is not the same for all phonons. To simplify matters, we assume that there is an average distance, and that all phonons travel this distance between one collision and the next. I call this distance the mean free path, and denote it by l. This means that on the average, a phonon carries its energy a distance l down the bar before it has a collision and gives up the energy to other phonons, and they in turn continue the process. Experiments have been done to measure *l*, and I show some results in table VIII-2*a*.

Table VIII-2*a*. *Mean free paths of phonons at room temperature*

material	mean free path in angstroms
quartz	100
silicon	430
common salt	70

The electrons and holes in a metal also suffer collisions, with phonons and with alloying atoms in the case of alloys. They therefore have a mean free path too, and some values are listed in table VIII-2*b*.

Table VIII-2*b*. *Mean free paths of electrons and holes at room temperature*

metal	mean free path in angstroms
copper	850
silver	1140
iron	440

The distance between neighbouring atoms in a solid is a few angstroms (one centimetre is equal to one hundred million angstroms). So you can see that all these particles go a fair distance compared to interatomic distances before they undergo collisions. If they were particles in the old-fashioned sense, one might expect their mean free path to be of the order of the interatomic distance. It is the quantum behaviour of the particles which is responsible for the long mean free paths.

4 Thermal conductivity of insulators

I now take a closer look at how thermal energy is carried down the bar by phonons, and use fig. VIII-4 for this purpose. Let the heat flow be from left to right, which means

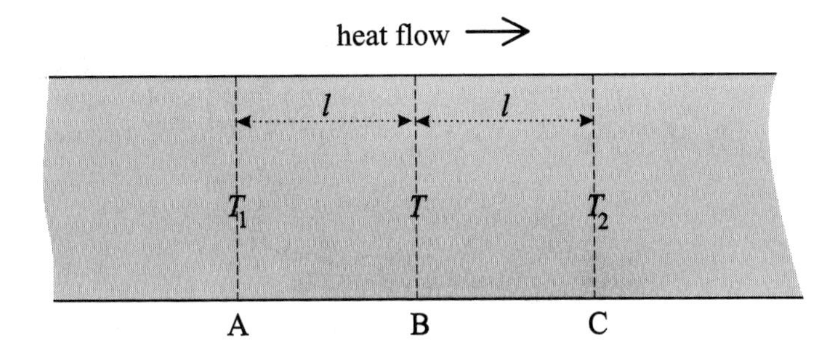

FIGURE VIII-4. A, B, and C are imaginary planes each of area *S* and spaced a mean free path *l* apart, cutting the bar along which heat is flowing. They are at temperatures T_1, T, and T_2 respectively, with T_1 the highest and T_2 the lowest.

that the temperature decreases as I go in this direction along the bar. Imagine three cross-sections A, B, and C of the bar which have the same area of *S* square centimetres, at a distance *l* apart, where *l* is the mean free path in centimetres. Suppose that the temperatures at A, B, and C are T_1, T, and T_2 respectively. Because of the direction of the heat flow, we know that T_1 is higher than T, and T is higher than T_2. The phonons at A have an energy $E(T_1)$ per cc, the thermal energy at temperature T_1. On the average, half of them are going to the right and reach B before they suffer collisions. How much energy per second flowing to the right do these phonons deliver at B? The problem is similar to the following: I stand at the side of a road on which pass cars, all at the same speed of 100 km per hour and so spaced apart that there are 10 cars per km of the road. How many cars do I see going past in one hour? The answer is 10 cars/km multiplied by 100 km/hour, or 1000 cars per hour. By a similar reasoning, the thermal energy flowing to the right per second at the surface B is given by

$\frac{1}{2}E(T_1)$ times the area *S* times the speed *v* of the phonons.

Note that the speed of the phonons is the same as the speed of sound, which in most solids is a few thousand metres per second. There are also phonons at B which are travelling to the left. These phonons arrived there from C where they had a thermal energy $E(T_2)$ per cubic cm, and they bring a thermal energy per second flowing to the left equal to

$\frac{1}{2}E(T_2)$ times the area *S* times the speed *v* of the phonons.

So the net thermal energy flowing to the right per second, which I denote by Q_L, is equal to the difference of the above two quantities:

$$Q_L = \text{net heat flow per second across B} = \tfrac{1}{2}[E(T_1) - E(T_2)]Sv.$$

I now do a little bit of arithmetic to get the above formula for Q_L into a form which shows me how to relate Q_L to the specific heat and the mean free path, which are properties of the phonons in the material. I multiply Q_L by two numbers which are each equal to one, so that the result is still Q_L:

$$Q_L = Q_L \text{ times } \frac{T_1-T_2}{T_1-T_2} \text{ times } \frac{2l}{2l}$$

Now notice that Q_L contains in it the factor $[E(T_1) - E(T_2)]$ which is just the difference in the thermal energies per cc between the temperatures T_1 and T_2. This difference divided by the difference in temperature is the specific heat of the material, namely the energy needed to raise the temperature of one cc of the material by one degree. I denote this as before by C_L:

$$Q_L = \tfrac{1}{2}SvC_L\frac{T_1-T_2}{2l}2l.$$

The temperature of the bar drops from T_1 and T_2 over a distance $2l$. The quantity $\frac{T_1-T_2}{2l}$ (which I denote by R) is the rate of change, or the gradient, of temperature along the bar. I can then write the amount of thermal energy going down the bar per second as

$$Q_L = vC_L lSR.$$

I have assumed so far that all the phonons are travelling either to the left or to the right along the bar. In fact, of course, they are travelling in all directions. When I take this also into account, the formula for Q_L is modified to

$$Q_L = \tfrac{1}{3}vC_L lSR,$$

which I can write as

$$Q_L = K_L SR.$$

I have used the symbol K_L to stand for $\tfrac{1}{3}vC_L l$. The quantity K_L is purely a property of the material; it depends on the specific heat C_L due to the phonons, their speed v, and their mean free path l. The subscript L reminds us that the heat is being carried by the *l*attice vibrations, i.e. the phonons. K_L is called the *lattice thermal conductivity* of the material. The larger it is, the better is the conduction of heat in the material.

This picture of how heat is carried down a bar by phonons allows us to understand the features of heat conduction mentioned in section 1: heat flow down a bar requires a difference in temperature between the two ends of the bar, and takes place rather slowly. The formula given above for Q_L shows that when R is zero, i.e. when there is no temperature difference, the heat flow is also zero. The rather slow speed with which heat travels down the spoon when one end is dipped in hot coffee is because of the collisions which the phonons suffer after travelling the distance of a mean free path which is very small in comparison to the length of a spoon. If a rod which is a single crystal is cooled to a temperature very near the absolute zero, then the mean free path for phonons becomes very large, and a pulse of heat fed in at one end travels to the other end at a speed comparable to the speed of sound in the material. One says then that the thermal energy is carried by second sound.

Now consider a bar of material which is in the shape of a cube, one cm on the side, as illustrated in fig. VIII-5. By connecting the cube to suitable warm plates and so on, one can arrange that the faces A and B are held at temperatures of 21 °C and 20 °C respectively. The bar has a cross-section of one square cm, and the temperature gradient is one degree per cm. From the formula for Q_L given above, we can see that the heat flowing per second from the face A to the face B is just K_L. Thus the thermal conductivity as we have defined it above is the thermal energy flowing per second between two opposite faces of a cube one cm on the side, when the temperature difference between the faces is one degree. I show in table VIII-3 this quantity in joules per second for various materials in the form of such a cube.

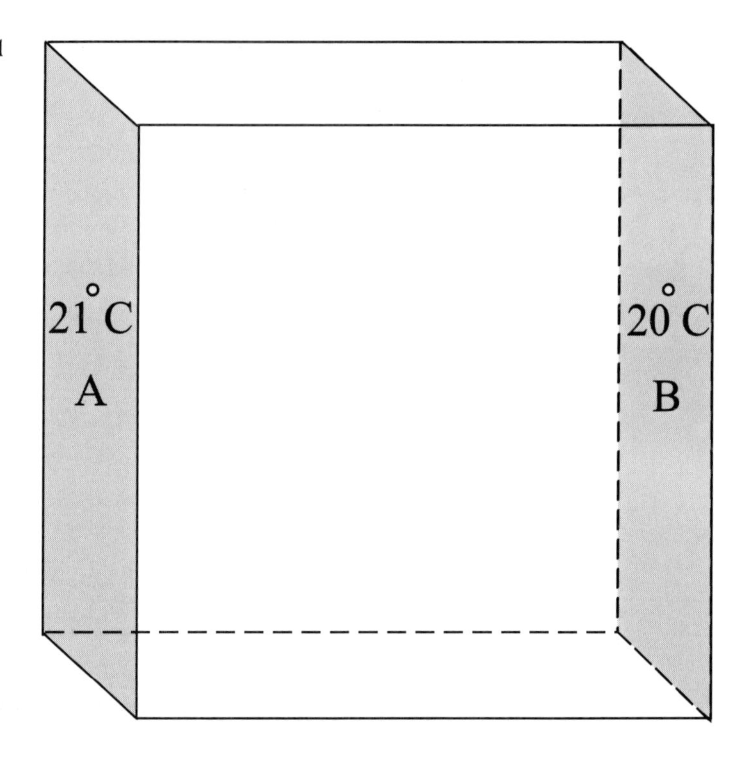

FIGURE VIII-5. The cube of material used to define its thermal conductivity. Its edges are one centimetre long, and the pair of opposite faces A and B are at 21 °C and 20 °C respectively.

Table VIII-3. *Flow of heat per second for the cube shown in fig. VIII-5*

material of cube	flow of heat in joules per second
quartz	0.07
glass	0.01
silicon	1.5
common salt	0.064
silver	4.2
copper	4.0
stainless steel	0.14
brass	0.8
typical plastic	0.1

A joule is the unit with which we measure amounts of energy. A joule of thermal energy added to one millilitre (which is one cc) of water raises its temperature by almost one quarter of a degree Celsius.

The numbers in table VIII-3 show some interesting regularities. Metals like copper and brass conduct heat better than insulators like glass and plastic, and a pure metal like copper is a better conductor of heat than an alloy like brass, which is a mixture of copper and zinc. The principal reason why metals conduct more heat than crystalline insulators under the same conditions is that, in addition to the phonons which are common to both kinds of materials, the metals have free electrons coming from the outermost shells of the atoms. These electrons can move about the material and make an additional contribution to the flow of thermal energy. We look at this electronic part of the thermal conduction in metals in the next section. We shall then be able to see why alloys do not conduct heat so well as pure metals.

5 Thermal conductivity of metals

The thermal energy in insulators and semiconductors is almost all contained in the phonons. In a metal on the other hand, the thermal energy is contained not only in the phonons, but also in the electron-hole pairs which appear near the Fermi surface because of temperature (fig. VI-18, p. 124). The flow of heat in a metal is therefore due to the phonons and also the electrons and holes.

The way that thermal energy is conducted by the electrons and holes is exactly the same as the way by phonons. I get the same formula for the thermal conductivity as in the case of the phonons, except that I must use in it the corresponding quantities for the electrons and holes: the specific heat C_E, the Fermi velocity v_F, and the electron's mean free path l_E. Calling this electronic thermal conductivity K_E, I get

$$K_E = \tfrac{1}{3} C_E v_F l_E.$$

So the thermal energy carried per second by the electrons and holes is

$$Q_E = K_E S R,$$

and the total thermal energy is Q, with

$$Q = Q_E + Q_L.$$

Notice that in each case the thermal conductivity is determined by the product of the specific heat, the speed and the mean free path of the particles in question.

We now estimate the relative magnitudes of the electron and the phonon contributions to the conduction of heat. We consider the mean free paths first. An electron or a hole collides with a phonon long before it meets another electron or hole. The most likely collisions for phonons also are with other phonons. The mean free paths for all of them therefore tend to be about the same, as can be seen in tables VIII-2*a* and VIII-2*b*. Although the electronic specific heat is smaller than the lattice specific heat, the Fermi velocity is much larger than the speed of sound in the material. The result is that the heat conduction due to electrons and holes is comparable to or even larger than that due to phonons.

In an alloy, we have more than one kind of atom in the lattice: copper and zinc atoms in brass, for example. This means that we have lost the translational symmetry of a pure metal in which every lattice point is occupied by the same kind of atom. This departure from perfection acts as a disturbance to the particle (electron, hole or phonon) in its quantum state. It responds to the disturbance by moving to a different quantum state, which is just a different way of saying that the particle is scattered. This additional scattering means that the thermal conductivity is decreased by alloying.

You will notice from table VIII-3 (p. 154) that quartz is a much better conductor of heat than glass, even though they both consist of molecules of silicon dioxide. So the mean free path for phonons in glass must be smaller than in quartz. The difference between the two materials is that quartz is a crystal and the molecules are arranged in it with translational symmetry, whereas in glass there is no such symmetry. So we come to the general conclusion that loss of translational symmetry, whether caused by alloying or by conversion into a glass, causes the mean free paths of phonons and electrons to decrease. As a result of this the thermal conductivity becomes less.

6 Thermal expansion

When a metal cap on a glass jar has been screwed on so tightly that it cannot be unscrewed by hand, an old kitchen trick is to heat the cap, for example by running hot water over it. This works in some cases, and the cap comes off easily. This could be because a bit of whatever was in the jar had gone hard in the space between cap and jar, and the heating simply softened it. But the method can also work when everything is clean. The reason then is that the heating causes the cap to expand in size so that it is no longer a tight fit on the jar. Such a change in size of a material when its temperature is changed is called its thermal expansion. In practically all situations that we are likely to encounter, it is extremely small and would not be seen by the unaided eye. Table VIII-4 shows the thermal expansion of some common materials.

Table VIII-4. *Thermal expansion at room temperature of some materials*

(The number in the second column gives the increase in length of a rod of the material one cm in length when its temperature is raised from 0 °C to 100 °C)

material	increase in length in cm
copper	17×10^{-4}
iron	12×10^{-4}
aluminium	24×10^{-4}
stainless steel	10×10^{-4}
glass	5×10^{-4}

You notice that different materials expand by different amounts for the same rise in temperature, and that glass expands less than the listed metals. One sees now why a stuck bottle cap can be freed by heating it, even if the glass gets a bit warm.

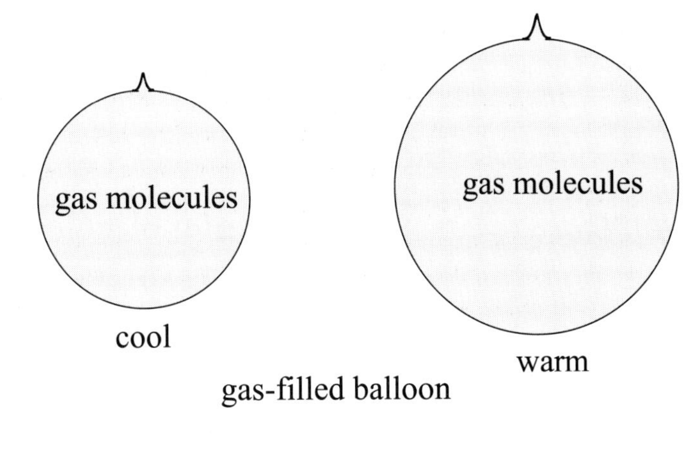

FIGURE VIII-6. The upper part shows the expansion of a warmed balloon, and the lower part that of a warmed solid. The molecules of the gas (the phonons, as also the free electrons if a metal) inside the balloon (in the solid) exert a pressure on the balloon wall (on the surface of the solid). The pressure goes up if the temperature goes up, and the balloon (the solid) expands a bit as a result.

What causes a solid to expand when it is heated? I take a simple example of something which shows similar behaviour, namely a rubber balloon filled with gas, shown in the upper part of fig. VIII-6. The pressure of the gas which is trying to expand the balloon is exactly balanced by the elastic inward pressure of the balloon which is trying to shrink it. It is this balance which determines how big the balloon will be for a given pressure of the gas inside it. When I warm up the balloon, the pressure of the gas increases, and this is counteracted by the balloon expanding and therefore increasing its inward pressure so as to continue to balance the gas pressure. The amount of gas in the balloon remains the same through all this.

Now imagine, as shown in the lower part of fig. VIII-6, the balloon replaced by the surface of the solid, and the gas molecules by the phonons in the solid, and the free electrons also if the solid is a metal. These phonons and electrons exert a pressure on the surface, and the pressure increases with increasing temperature. This results in the thermal expansion of the solid.

So far, so good: but how are we to understand the pressure in terms of the molecules in a gas, or the phonons and electrons in a solid? I take the case of a gas first. Figure VIII-7 shows a portion of the balloon wall and the paths of some of the many molecules

FIGURE VIII-7. Molecules of a gas bouncing off the containing wall (of a balloon, for example) transfer momentum to the wall. The amount of momentum so transferred per second is proportional to the pressure on the wall.

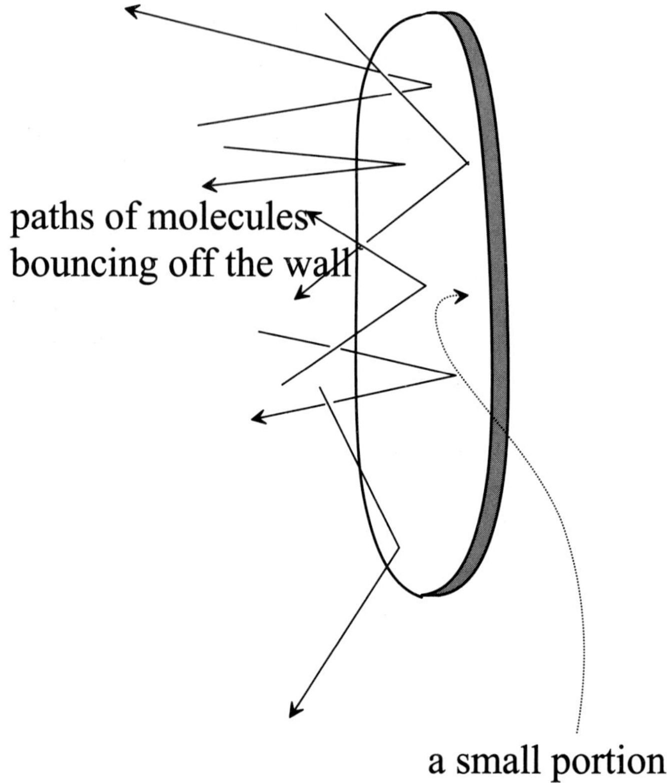

that are hitting it and bouncing back all the time. As each molecule bounces back, it causes a recoil of the wall directed to the right in the figure. What is at work here is the conservation of momentum. The momentum of the molecule changes as a result of the collision with the wall. This change must be compensated by a change of momentum of the wall, such that the total momentum remains unaltered.

The cumulative effect of such recoils due to all the molecules hitting the wall is that the gas pressure on that side of the wall tends to move it to the right. At the same time, the wall suffers similar collisions from molecules on its other side too, and these give it an impulse which tends to move it to the left. The impulse, and therefore the gas pressure, increases when the speeds of the molecules increase, which is what happens when the temperature of the gas rises. Thus we see why the balloon expands when the gas inside it gets warmer.

QUESTION Where does the elasticity of the balloon come into all this?

ANSWER The excess pressure of the warm gas inside the balloon over that of the cool gas outside is balanced by an effective pressure of the balloon acting inwards, resulting from the fact that it has increased in size. This is a bit like a stretched rubber band trying to return to its original size.

The free electrons in a metal, which are largely responsible for its thermal conductivity, are moving about inside the metal, but cannot get out through its surface. The attraction between the negatively charged electrons and the positively charged ions in the metal holds the electrons trapped inside the metal. The effect of this attraction is as if there were an impermeable but elastic sheath all over the surface of a piece of metal. The electrons are confined inside this sheath, with which they are constantly colliding and recoiling, in analogy with the gas molecules and the balloon wall. There is a resulting pressure due to the electrons on the surface of the metal. The average speed of the electrons increases when the temperature rises, thus causing this pressure to increase and the metal to expand. We call this the electronic contribution to the thermal expansion.

Now how about the phonons? The electrons have mass, and they are moving about at various speeds. It is easy to imagine that they transfer impulse to the wall when they collide with it. The picture is a little different with phonons. They represent the quanta of energy of the vibrating atoms, and there is no obvious way in which we can associate a mass with a phonon. A phonon carries no momentum in the sense in which a molecule in a gas does. You can see this when you recall that the phonon is the quantum aspect of a sound wave in the solid. For each atom in the sound wave with a given speed in some direction at any given instant, there is another atom with the same speed but in the opposite direction, so that the total momentum of the wave will be zero. We cannot talk about a phonon colliding with the surface of the solid and transferring momentum to it.

There is a different way of looking at things which gives meaning to the idea of the pressure due to phonons. To begin with, the frequency, and thus the energy, of a phonon decreases when the volume of the crystal increases. If there is nothing to hold the crystal together, this would mean that even when the temperature is constant, the crystal would keep expanding because its energy due to the phonons would then keep getting lower. One can think of this as a pressure inside the crystal caused by phonons which tends to expand it. The crystal of course does nothing of the sort, because there are internal forces due to the charged nuclei and electrons which hold the crystal together. If now the crystal is heated up a bit, the distribution of phonons in the different energy levels changes, and consequently the phonon pressure also changes. This results in a change in volume of the crystal.

This way of relating thermal expansion to the change in quantum energy levels produced by a change in volume is an example of how quantum mechanics and statistical mechanics are combined. It can also be used to get the contribution of the conduction electrons in a metal to its thermal expansion.

7 Some practical examples

We have looked at three properties of materials which are related to temperature: specific heat, thermal conductivity and thermal expansion. I give some examples of everyday situations where these properties come into play.

While eating a meal, you may have on occasion put a potato in your mouth which was much hotter than you expected. You may then have noticed how even a sip of water is enough to cool off the potato. This is because the specific heat of water is much greater than that of the potato. Even though the amount of water is small, it undergoes only a modest rise in temperature as it absorbs the thermal energy coming out of the potato as it cools. The specific heat of water is one of the highest for any material we know, as is evident in table VIII-1. This is why, for example, if a pan of water and a garden tool are left outside in sunlight, the tool may become too hot to touch whereas the water may be barely warm. It is this ability of water to absorb large amounts of thermal energy with only modest changes in its temperature which is an important factor in determining the earth's climate.

In solid materials, thermal energy is carried by the process of conduction which I have described here: energy moves, matter stays where it is. In liquids and gases, on the other hand, the process is predominantly *convection*: matter itself moves, carrying energy with it. There is still a third way in which thermal energy moves, namely as *radiation*, and I shall say more about it in the next chapter. I consider here a common situation in which thermal conduction plays a central role: household heating in the winter months, or cooling in the summer months.

Figure VIII-8 is a bare-bones impression of a house in winter. The outside temperature is 5 °C and the inside is heated to hold a temperature of 22 °C. Since the inside temperature is constant, there can be no energy being absorbed there. Thus all of the energy being put out by the heating system finds its way to the outside world by conduction through the walls, roof, and other boundaries of the house. Once the inside temperature has been brought up to its steady level, all the energy supplied by the heating system is warming up the world outside the house. This is of course the reason for using very good thermal insulation between the inside and outside of the house; one wants to reduce the thermal energy which is being conducted to the outside. We saw earlier (p. 152) that the rate of energy flow is proportional to the difference in temperature which produces this flow.

QUESTION What is the saving in energy if the house temperature is set at 18 °C instead of at 22 °C in the example above?

ANSWER The energy flow from inside to outside is proportional to the difference of temperature between the two. When the setting is changed from 22 °C to 18 °C, this difference changes from 17 degrees to 13 degrees. Thus the heating power now needed will be 13/17, which is 72 per cent, of what was needed before.

Another example of the use of our knowledge about thermal conductivity is found in the kitchen: stainless steel cooking pots with copper bottoms. You can see from table VIII-3 (p. 154) that copper is thirty times as good a conductor of heat as stainless steel. The copper bottom spreads the heat from the stove uniformly and therefore helps to prevent hot spots with their deleterious effect on what is in the pot.

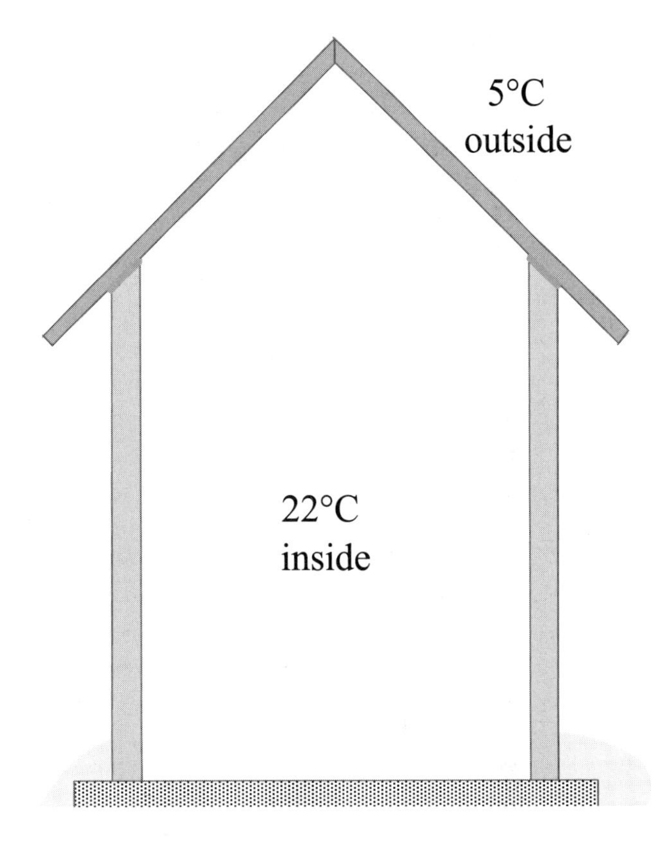

FIGURE VIII-8. Impression of a heated house in winter: 5 °C outside and 22 °C inside. All the thermal energy supplied by the heating system, once the inside temperature has stabilised, flows through the floor, walls and roof of the house to heat up the outside world.

The thermal expansion of materials between the highest and lowest temperatures that we encounter around us is quite small. For example, a rod of iron one metre long would increase in length by about one millimetre between minus 50 °C and plus 50 °C. Although this effect is small enough to be negligible for many purposes, it has to be explicitly taken into account in the design of many outdoor structures: railway tracks, roads, bridges, and buildings, for example.

8 Summary

I apply in this chapter the quantum mechanical picture of a solid to the thermal properties of a solid. One of these is the conduction of heat: section 1 considers some examples and deduces some general features of how thermal energy flows down a rod. A difference of temperature must exist between the ends of the rod, and the flow is from the hot end to the cool end and is rather slow.

The specific heat of a solid is one of its thermal properties which is not only of intrinsic interest, but also important for the thermal conduction. It is a measure of how much energy must be fed into a solid in order to raise its temperature by a given amount. Section 2 describes the contribution to the specific heat from phonons (which exists for all solids) and from free electrons (which applies to metals only).

The carriers of the energy in thermal conduction are phonons, electrons, and holes. Their motion in the solid is not unhindered: they undergo collisions which can change their energy and momentum. We can think about the average distance such a particle travels between collisions, and call it the mean free path. This is the subject of section 3.

The ideas from sections 2 and 3 are brought together in section 4 to give a picture of how thermal energy is conducted down an insulator by phonons. We see how the specific heat and the mean free path are involved in thermal conduction.

Section 5 takes up thermal conduction in metals. Here the contribution from free electrons is added to that from phonons. This explains why in general metals conduct heat better than insulators. Since the atomic arrangement in alloys is less ordered than in pure metals, the mean free paths for electrons as well as for phonons are smaller in alloys. This is the principal reason why, for example, stainless steel is a poor conductor of heat compared to copper.

Section 6 deals with thermal expansion, which is the change in size of a piece of matter when its temperature changes. This arises from the fact that the quantum states of the electrons and phonons are modified when the volume of the piece changes.

Some examples from everyday experience are mentioned in section 7 to illustrate the thermal effects discussed in this chapter.

This brings us to the end of our consideration of heat and matter. In the next chapter, I shall describe to you how light interacts with matter. We go on to where there will be more light and less heat.

IX GLASS PANES AND ALUMINIUM FOILS

1 Some preliminaries

A sheet of glass is transparent to light, whereas a much thinner sheet of aluminium foil is opaque: materials can be divided into those which let light go through, and others which block it. Examples of the first kind are glass, clear plastics, crystals like sugar and common salt. Metals on the other hand are opaque to light. We know that all these materials are made up of atoms. We saw in chapter IV that the atom itself is mostly empty space, because its nucleus and electrons occupy a tiny fraction of the volume of the atom. We might think therefore that light, which can travel freely through empty space, should be able to go right through all materials, but this clearly is not so. The transparency or opacity of materials results from what happens to the quantum energy states of the electrons when the atoms come together to form a solid.

Colour is another property of materials which must have something to do with how light interacts with them. Metallic copper for example is red and gold is yellow, in daylight. Sapphire is blue and ruby is red. The explanation again lies in the electronic structure of such solids, together with the quantum nature of light, namely that it consists of photons with energy equal to the frequency multiplied by the Planck constant.

The kind of wave motion, of which visible light is an example, extends over a tremendous range of frequencies. Radio waves, microwaves, and X-rays are the same kind of waves as light, differing only in frequency. These waves collectively are known as *electromagnetic waves*, for reasons which will become clear later. We shall find out what an electromagnetic wave is, and then look at some examples of how these waves interact with matter. I shall consider here mainly visible light, because this is the kind of electromagnetic wave that we see every day. The picture that I shall develop applies to all the other waves too.

2 What is light?

In chapter VII, I described heat as a form of energy which a piece of material has by virtue of its temperature. This thermal energy was seen to be the energy of motion of the atoms and electrons in the material. We cannot therefore imagine this kind of energy in the absence of matter.

There is another example of energy, namely the energy we receive from the sun in the form of light. Sunlight reaches us after travelling a path from the sun which is mostly empty space, and can go through materials like window glass but is stopped by other materials like aluminium foil. We have seen in chapter III that light has the property of duality: it behaves like waves showing interference and diffraction, and like particles (which we call photons) in the effect discovered by Arthur H. Compton when a beam of X-rays is scattered by electrons. The momentum $\mathbf{p}$ and the energy E of a photon are related to the wave vector $\mathbf{k}$ and the frequency f of the corresponding light wave by the rules of quantum mechanics:

$$\mathbf{p} = h\mathbf{k}$$

and

$$E = hf,$$

where h is the Planck constant. The waves which correspond to visible light have wavelengths between about 8×10^{-5} cm for red light and 4×10^{-5} cm for violet light. The other colours have wavelengths between these two values. The frequency is given by the speed of light, 3×10^{10} cm/sec, divided by the wavelength.

In a wave on the surface of water, it is the water itself which is moving up and down. In a sound wave travelling through air, the pressure at each point oscillates up and down at the frequency of the sound wave. In both these waves, there is a material, water or air, which carries the wave and oscillates in a known manner. But what about a light wave? It can travel through empty space, and therefore does not need a material medium which does the oscillating. It goes through a vacuum and also through certain, but not all, materials. In order to see why glass is transparent to light, whereas aluminium foil is not, we need to look more closely at exactly what it is that is oscillating in a beam of light. Before doing that, I would like to tell you something about the wavelength and frequency of these waves.

Visible light has a wavelength, depending on its colour, between 4×10^{-5} cm and 8×10^{-5} cm. There also are waves with wavelengths outside this range, not seen by our eyes, but otherwise exactly like light waves. Such waves exist in nature, or can be produced with suitable instruments. Their wavelengths have been observed to range from 10^{10} cm all the way down to 10^{-14} cm. There can also exist such waves with wavelengths beyond these limits.They all travel through vacuum with the speed of visible light, 3×10^{10} cm per second. These waves, all having the same basic character and spanning this vast range of wavelengths and frequencies, are collectively known as *electromagnetic radiation.*

Figure IX-1 shows the spectrum of electromagnetic radiation. It is divided into different regions whose approximate boundaries are shown by broken lines. The waves in each region have their own name: gamma rays, ultraviolet light and so on. The ways of producing the waves are different for the different regions. For example, gamma rays are generated by radioactive nuclei, while microwaves are produced using special vacuum tubes. Frequencies and wavelengths belonging together are connected horizontally in the figure: gamma rays of frequency 10^{20} hertz (point A) have a wavelength of 3×10^{-10} cm (point B), for example. Since the speed is equal to the frequency multiplied by the wavelength, these gamma rays travel at 3×10^{10} cm per second, which is

FIGURE IX-1. The different regions of the electromagnetic spectrum. Note the scales showing frequency and wavelength: each step denotes a change by a factor of 10. Such scales, called logarithmic scales, are used when the numbers vary over a tremendous range of values, as is the case here. 1 hertz = 1 oscillation per sec.

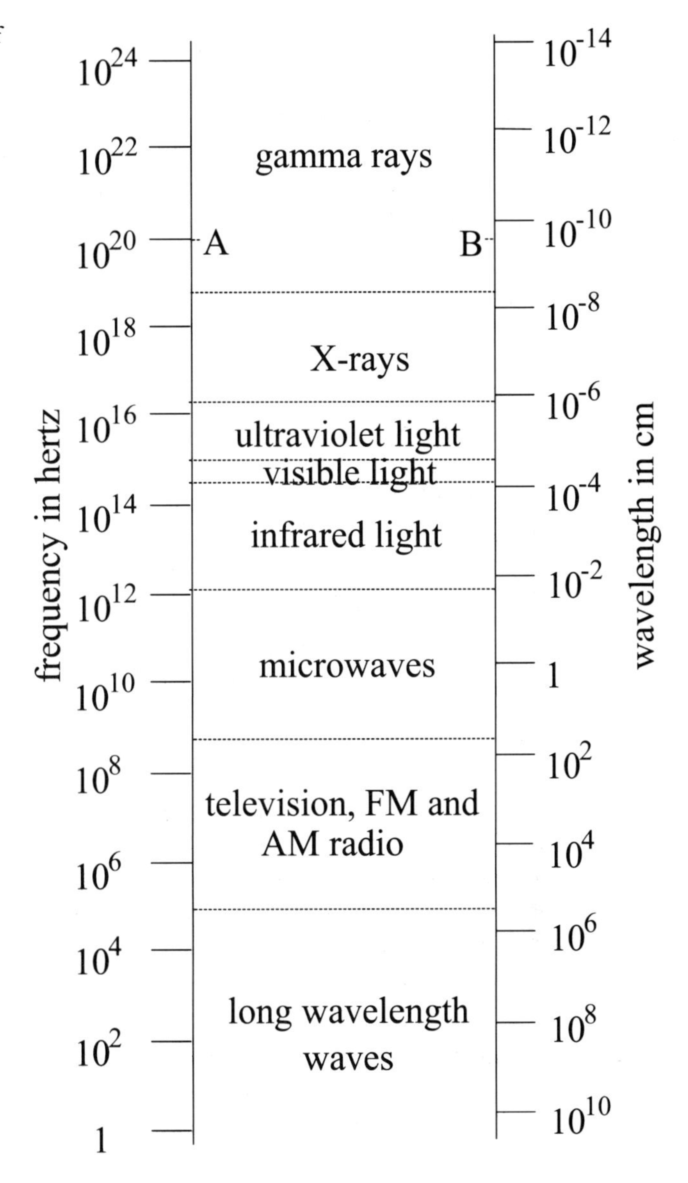

also the speed of visible light. You can verify from the figure that this is the speed of all the waves in the entire spectrum.

3 Electric and magnetic fields

In this section, I describe what it is that is waving in an electromagnetic wave. In order to do this, I must first explain what is meant by the term *field* as used in physics. This is

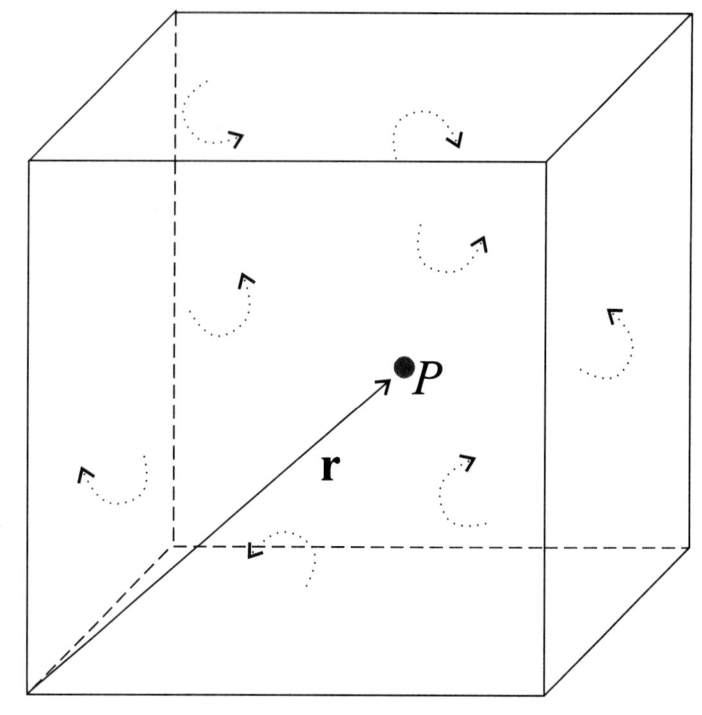

FIGURE IX-2. The arrows suggest the motion of different parts of the air inside a heated oven. The temperature of the air and the direction and magnitude of its speed vary from point to point. One speaks of temperature and velocity fields for the air in the oven.

a term which we use here with a special meaning which is different from what it has in everyday speech. I begin with a simple example: a kitchen oven which has been turned on. I show the oven in fig. IX-2 as a hollow box. The air inside the oven is steadily getting hotter. At some instant of time, which I denote by the symbol t, I measure the temperature T at every point inside the oven. I can specify any point P by noting its position vector $\mathbf{r}$ from one corner of the oven, as shown in the figure. Changing the direction and/or magnitude of $\mathbf{r}$ will let me place P anywhere in the oven. What I have done, at the chosen instant of time, is to make a snapshot picture, so to speak, of the temperature everywhere inside the oven. I represent this picture by the symbol $T(\mathbf{r}, t)$. I choose this form for the symbol to be reminded of the following:

1. It stands for the temperature T.
2. The temperature depends on the position $\mathbf{r}$ and time t.

This symbol stands for the temperature at every point in the oven at all different times, and is therefore a very convenient shorthand for a lot of information. In words, I say that *there is a temperature field inside the oven, which depends on position* $\mathbf{r}$ *and time t, and is denoted by* $T(\mathbf{r}, t)$.

QUESTION Imagine that the air inside the oven is not still, but moving about as suggested by the curved arrows in fig. IX-2. Can you think of a field associated with this motion?

ANSWER Yes. At each point P the air is moving with a certain speed in some direction, both of which may change with time and with the position of P. I can therefore think of a velocity field which I can denote by $\mathbf{v}(\mathbf{r}, t)$ for the air in the oven.

The temperature field is a quantity which has a magnitude (so many degrees Celsius, for example), but no direction. We call such a quantity a *scalar*. The temperature inside the oven forms a *scalar field*. The velocity of the regions of air inside the oven forms a *vector field*. What is common to both fields is that, over a given volume of space and span of time, a certain quantity which can be a scalar (e.g. temperature) or a vector (e.g. velocity) is specified at every point.

The two fields I have described, temperature and velocity, are properties of the medium, namely air, which fills the space of interest, namely the oven. I now look at another example which will present a new and very important aspect of the field. Consider an electric charge, say an electron, which is fixed at some point A in space. If I now place a unit positive electric charge at a point B anywhere in the space surrounding A, it will feel a force because of the electron's charge. Figure IX-3 shows how this force decreases as the distance between A and B increases. The force is a vector, since it has both a magnitude and a direction, and it can be measured at all points in the space surrounding the electron A. Thus this force satisfies the conditions for a vector field. I give this force a name and a symbol: the *electric field due to an electron*, with the symbol $\mathbf{E}(\mathbf{r}, t)$. The vector $\mathbf{r}$ here is the vector from A to B, and specifies the point in space at which I measure the electric field at time t. I shall sometimes denote the electric field more simply with just the symbol $\mathbf{E}$, keeping in mind that it may vary with time and from point to point.

I can write down a formula for the force which the unit positive charge feels due to the charge of the electron. I just have to multiply the charges together and divide by the square of the distance apart:

$$\mathbf{E} = 1 \times e/r^2 = e/r^2 \text{ in magnitude, pointing towards the electron.}$$

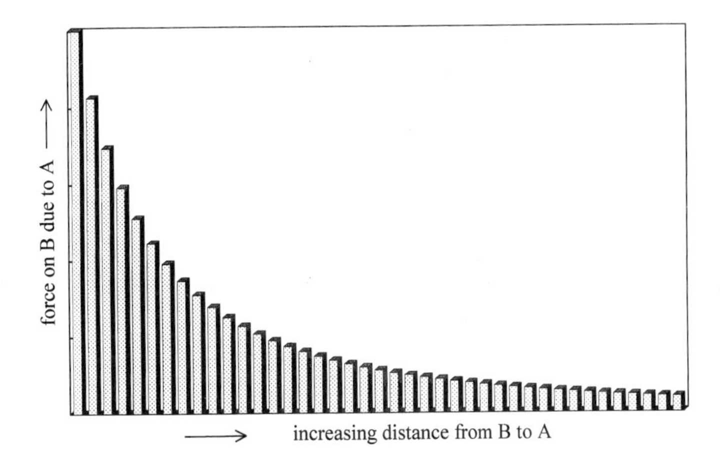

FIGURE IX-3. The force on a unit positive electric charge placed at B due to an electron with charge $-e$ placed at A, as the distance between the two is varied; the force decreases as the distance increases.

Here e is the magnitude of the charge of the electron. The direction of the field is as indicated because unlike charges attract each other.

Now imagine that I replace the unit positive charge at B with some electric charge which I denote by q. Then the force on this charge is given by

$$qe/r^2$$

towards or away from the electron depending on whether q is positive or negative. I can include all of this in the statement that the force vector on q is equal to $q\mathbf{E}$ whatever value q takes. It is as if the presence of the electron at A influences the properties of the space around it such that any charge placed in it feels a force determined by the electric field due to the electron at its location. I take one more step in my thinking, and say that the presence of the electron at A creates an electric field at B, where B is any point in the space around A. In saying this, I am going beyond treating the electric field as merely a way of thinking about the force on the charge at B due to the electron at A. I imagine that a real physical entity, which I call the electric field, exists at B whether there is or not an electric charge at B. All that is necessary for the field is that there should be somewhere a charge or charges to act as the source of the field. By focusing my attention on the field, I move away from thinking about the charge which is needed to produce it in the first place. I can look at a given volume of space, even if it is free of charges, and imagine an electric field in it, which in general could vary with time, without explicitly worrying about the charges which must be somewhere in order to have produced the field.

Now I come to the magnetic field. A compass needle is a simple example of a magnet. It is a bar magnet which can rotate horizontally about a pivot. Its north pole points to the geographic north pole, because the earth has magnetic properties that

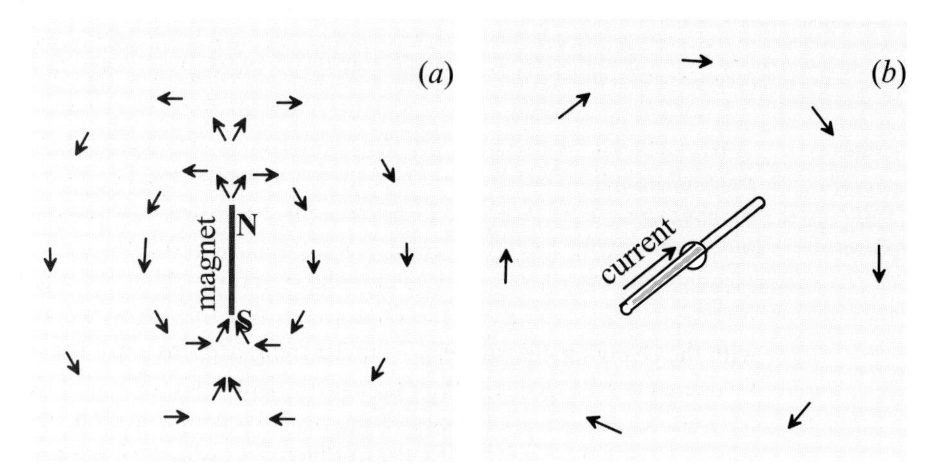

FIGURE IX-4. How compass needles, shown as arrows, orient themselves at different points: (*a*) around a bar magnet, and (*b*) around a wire carrying a constant electric current. In both cases, the arrows indicate the direction of the magnetic field.

resemble those of a large bar magnet lying along its axis of rotation. I illustrate in part (*a*) of fig. IX-4 the influence of a bar magnet on compass needles placed in various positions in its neighbourhood. At each position, there is a definite direction in which the needle points. To turn the needle away from this direction requires a twisting torque which varies from point to point, becoming smaller the further it is from the magnet. Thus the influence of the bar magnet upon the compass needle has two aspects: a direction (the direction of the needle) and a magnitude (proportional to the torque needed to disturb the needle). Further, it is an influence which pervades the space surrounding the magnet. I can include all of this in the statement that *a bar magnet creates a vector field in the space around it, called a magnetic field and denoted by the symbol* **B**. As with the electric field **E**, the magnetic field **B** may in general vary with time and position. The direction of the magnetic field vector determines the direction of the compass needle, and its magnitude tells us how strongly the needle sticks to this direction.

A bar magnet is not the only source for a magnetic field. An electric current flowing in a wire will also produce a magnetic field in the surrounding space, as illustrated in part (*b*) of fig. IX-4. It shows a short section of the current-carrying wire, and the orientation of compass needles placed at different points around the wire. The orientation shows the direction of the magnetic field, whose strength decreases with increasing distance from the wire.

I summarize now what I have said so far in this section. A *field* is a quantity which may be a scalar or a vector, whose value is specified everywhere in some volume of space and at different times. An *electric field* is produced by electric charges, and can be sensed by the force it exerts on an electric charge. A *magnetic field* is produced by a bar magnet or by an electric current, and makes itself known by its effect on a compass needle. Both these fields are vector fields.

Let us now imagine some volume of space which is empty of matter but has both an electric and a magnetic field everywhere in it. To begin with, let us assume that these fields do not change with time, although they may be different at different points in the volume. They are produced by some distribution of charges, currents, and magnets all lying outside the volume. We need to know nothing about them other than that they exist: our attention is focused on the field distribution inside the volume.

Now suppose something is done to the charges which are responsible for the electric field so that the field begins to change with time. We find a new phenomenon: the changing electric field produces a magnetic field which has nothing to do with the magnets and current-carrying wires outside the volume. Its connection is only with the changing electric field. A similar thing happens when the roles of the two fields are interchanged: a changing magnetic field produces an electric field. You can see some sort of symmetry between electric and magnetic fields here. A change with time of either field produces the other. The exact rules which govern this behaviour were written down in mathematical form by James Clerk Maxwell. The symmetry between the two fields was shown by Albert Einstein to follow from the theory of relativity.

A simple way to produce a magnetic field changing with time is to make the electric current in a wire oscillate back and forth in direction at a frequency of f oscillations per second. This means that the current increases to a maximum, drops to zero, changes direction and increases to a maximum, drops to zero, and so on, f times per second. The current executes what are called *sinusoidal oscillations*. Then the magnetic field at each point will also show sinusoidal oscillations with time. This oscillating magnetic field will, following Maxwell, produce an oscillating electric field, and so on. The oscillations of each field determine the oscillations of the other. These oscillations travel outwards away from the wire, with a speed which is set by the relation between the two oscillating fields in empty space (and is therefore a property of empty space): this is the speed of light. The experiment just described is in essence a way to produce a radio wave. We now see why such waves are called *electromagnetic waves*: the things which are undergoing wave motion are electric and magnetic fields. The entire spectrum shown in fig. IX-1 consists of such electromagnetic waves, all travelling at the same speed of 3×10^{10} cm per second, differing from one another only in frequency and corresponding wavelength.

Figure IX-5 shows a snapshot, as it were, of a portion of an electromagnetic wave. There are three significant directions which are associated with the wave: the direction along which the wave travels (shown by the wave vector **k**), and the directions along which the electric and magnetic fields point. These three directions are perpendicular to each other, as shown in the inset. This wave is somewhat different from the usual waves one sees, such as those on water. So let me describe how to interpret the figure. To be specific, suppose that it represents a wave of blue light travelling in the direction shown. Its wavelength is then about 4×10^{-5} cm.

As time goes on, the wave oscillates and travels in the direction of the wave vector **k** with the speed of light. I focus my attention on the point marked P as the wave passes

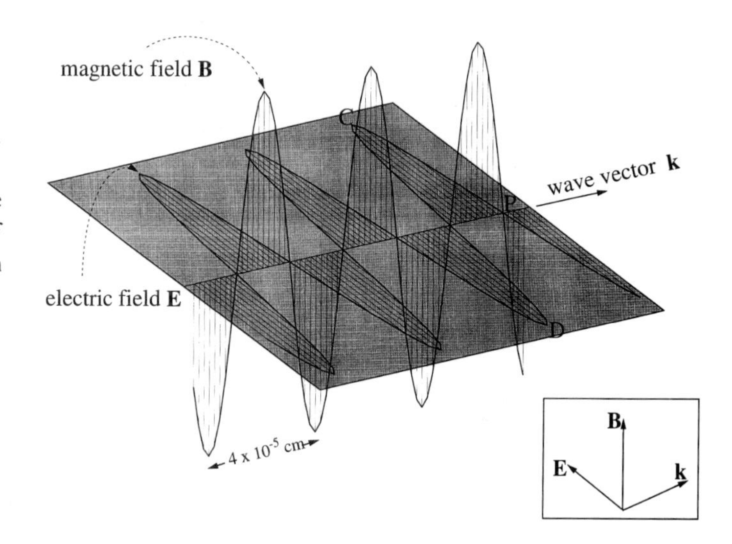

FIGURE IX-5. An electromagnetic wave of blue light travelling in the direction shown by the wave vector. The waves of electric and magnetic fields are coupled to each other. The mutually perpendicular directions of the vectors **E**, **B**, and **k** are shown in the lower right hand corner.

it. The wave consists of electric and magnetic fields, and to sense them I must place something suitable at P. To see what the electric field is doing, I imagine an electron at P. It senses the oscillating electric field of the wave moving past it. It feels an increasing force which reaches a maximum when the crest of the wave (point C in the figure) reaches it. The force then begins to decrease to zero and reaches a maximum in the opposite direction when the point D of the wave reaches P, and so on. Thus the electron feels an oscillating force at a frequency f, the frequency of the wave. The electron also has a magnetic moment, and so it senses the magnetic field too. Consequently, it also feels a force tending to align it along the magnetic field, oscillating at a frequency f.

You see from fig. IX-5 that the oscillating electric field remains in a fixed plane, So also does the magnetic field, but in a plane which is perpendicular to that of the electric field. This is so for all electromagnetic waves. If I look at the bundle of waves which make up a beam of light coming from the sun or from a lamp, I find that these two planes, while still perpendicular to each other, are differently oriented for the different waves. Such light is called *unpolarized light.* By passing this light through certain materials generally known as *polarizers,* it is possible to filter out all waves except those having the electric field in a single specified plane. The resulting light is called *polarized light.* A material which is often used in sunglasses contains layers of certain long molecules packed parallel to one another. Their electronic structure is such that they absorb waves whose electric fields are parallel to their length, and let through those whose fields are perpendicular to their length. The light reaching us from the sky is sunlight which has been scattered by the molecules in the atmosphere, and is partially polarized as a consequence. You can verify this by looking at the sky through polarizing sunglasses, and noting the change in brightness as you rotate the glasses one way and the other.

We have achieved a very powerful simplification by thinking of an electromagnetic wave in terms of its electric and magnetic fields. The wave of course needs some source to produce it: a light bulb if it is in the visible spectrum, a radioactive nucleus if it is a gamma ray, and so on. We do not need to worry about the complicated processes going on in the source which lead to the production of the wave in order to understand what happens when the wave *meets* a solid. Nevertheless, I describe briefly in the next section two ways in which electromagnetic waves can be generated from matter: this is where the wave *leaves* the source.

4 Generation of light

You have seen neon tubes which give off red light and are used in advertising signs. Such a tube contains neon gas, and is so constructed that when an electric voltage is applied between its ends, a current of electrons passes through the gas. These electrons collide with the neon atoms and transfer some of their kinetic energy to the atoms, which therefore change to one of their higher quantum energy states. The atoms subsequently lose this extra energy by emitting a photon and return to their previous state, and the process begins all over again. Thus the photons can only have those energies (and the corresponding frequencies) which are the differences between the higher energy states and the initial state of the atoms. In the case of neon, most of the photons come from a state whose energy is higher than the initial state by an amount which is the energy of photons of the colour red. With other gases, other colours may dominate the light that comes out. A characteristic of such light is that it consists of waves of only certain frequencies, which are fixed by the quantum energy levels of the atoms.

I can also produce light from a body if I make it hot enough: examples are the light from an incandescent electric bulb in which a filament of tungsten is made white-hot by passing an electric current through it, or from the sun which stays hot because of processes going on in it involving nuclei. When I pass such light through a glass prism, which bends light of different frequencies by different amounts, I get a continuous spread of colours from violet to red. By using suitable instruments, I find that the light also contains frequencies which the human eye cannot see, such as infrared (frequencies below the red) and ultraviolet (frequencies above the violet). In fact, the entire electromagnetic spectrum is present in such light, though most of the photons in sunlight are in the range of visible light, and there are fewer photons the further away one goes from this range.

How does this continuous range of frequencies come about? We know that the thermal energy of a body is just the energy of motion of its constituent particles, which are atoms, nuclei, electrons and so forth. The energies of individual particles vary from zero to very large values, and the distribution of the particles among the different energy ranges depends on the temperature of the body. We saw all this in chapter VI. It is this very complicated thermal motion of the electric charges, namely electrons and nuclei, which produces the continuous spectrum of electromagnetic radiation, which is therefore called *thermal radiation.* We know that if the temperature of the body is T kelvin, then the average energy of motion per particle is about $k_B T$, where k_B is the Boltzmann constant. I shall refer to this quantity as the thermal energy per particle. If I express the photon energy as so many times the thermal energy, and look at the distribution of the photons in the thermal radiation among different energy ranges, I get the result shown in fig. IX-6. The shape of this distribution results from the fact that photons are bosons, and is the same whatever the temperature of the body: 3000 K for the filament in an electric bulb, 6000 K for the surface of the sun, and so on. You can see that the largest number of photons are at energies which are about three times the thermal energy. As the frequency of the photon is proportional to its energy, this means that, for example, the largest numbers of photons in sunlight are at frequencies which are twice those in the light from an electric bulb. When I put the actual numbers in, I find that most photons from the sun are in the red part of the spectrum with a wavelength of about 8×10^{-5} cm, while they are in the infrared (wavelengths around 1.6×10^{-4} cm) in an electric lamp.

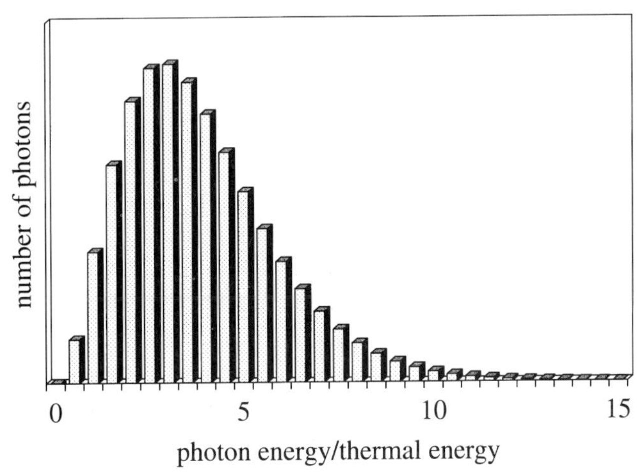

FIGURE IX-6. The energy distribution in thermal radiation, showing the relative numbers of photons in different energy ranges at any temperature T. The horizontal scale shows the ratio of photon energy to the thermal energy $k_B T$. While photons of all frequencies are produced whatever the temperature of the body, most of them are in a range of energies which are about $3k_B T$. This range is in the visible and infrared for light from the sun, whose surface temperature is about 6000 K.

QUESTION As you read this, you and your surroundings are also giving off a continuous spectrum of electromagnetic radiation. What frequencies do most of these photons have relative to those from an electric lamp?

ANSWER The temperature around you is about 20 °C which is the same as 293 K, which in turn is about one-tenth of the temperature of the filament in the electric bulb. So the radiation has most of its photons at frequencies which are one-tenth of those from the bulb. One can call this the far infrared part of the spectrum. It is this radiation which is picked up by devices used for night vision.

One can see from fig. IX-6 that the lower the temperature, the lower the range of photon energies at which the largest numbers of photons are found. The universe is filled with electromagnetic radiation which corresponds to a temperature of about 3 K; this means that the largest number of photons in this radiation have wavelengths in the vicinity of a millimetre. This radiation is a residue of the big bang which was the start of our universe 10^{10} years ago.

5 Absorption of light

All solids consist of particles, electrons and nuclei, which have two properties which make them feel the influence of the electric and magnetic fields in an electromagnetic wave: they are electrically charged, and they are permanent magnets. In addition, quantum mechanics tells us that the energies of the electrons are to be described in terms of energy bands, and the energy of the electromagnetic wave is quantized in terms of photons. Using this information, we now seek answers to questions like why glass lets light through and aluminium foil does not, and why copper is red and sapphire is blue.

In this section, I concentrate on the effects the electric field in a light wave has on the atoms which make up a piece of matter. The electric field acts on the electric charges, namely the electrons and nuclei which make up the atoms. The field acting on a given atom will push the nucleus (positive charge) in one direction and the surrounding cloud of electrons (negative charge) in the opposite direction. This relative displacement of the two stops when the force due to the electric field is balanced by the attractive force between the nucleus and the electron cloud. I shall come back to this effect later.

The light also behaves like photons, quanta of energy. A photon can be absorbed by an electron, whose energy is then increased by the energy of the absorbed photon. For this absorption to be possible, there must be a quantum state of this higher energy into which the electron can move. If there is no such state, then the photon cannot be absorbed and will therefore continue on its way. Now we begin to see why some materials are transparent and some are opaque.

Take glass as an example of a transparent material. It is an insulator, and we know qualitatively what its electronic structure in terms of energy bands must be. I show the two energy bands which are of interest for the present purpose in fig. IX-7. The lower energy band is full; every quantum state in it is occupied by an electron. Above it and separated by an energy gap is an empty band with no electrons. Lengths measured vertically on this figure represent energies. An incoming photon with energy represented by the line AB, which is less than the energy gap, cannot be absorbed by any of the electrons in the filled band, because the resulting final state will either be in the filled band and therefore already occupied, or must be in the energy gap where there are no quantum states for the electron at all. So such a photon would pass through the material. Now consider photons of energy larger than the gap energy, CD for example. An electron can absorb such a photon and go to a previously empty state in the upper band.

We see now why glass is transparent to visible light. The band structure of glass is such that the photons of visible light do not have enough energy to make an electron cross the energy gap, and so they pass through without being absorbed. But photons of ultraviolet light have higher frequencies and therefore higher energies, and these are enough to raise an electron from the full band to the empty band, so that such light would be absorbed by the glass. In the figure, AB would be a typical photon of visible light and CD of ultraviolet light. The energy gap in quartz is larger than in ordinary glass, and not only photons of visible light, but even those of ultraviolet light, have insufficient energy to raise an electron to a state across the gap. This is the reason that ultraviolet lamps are made of quartz and not of ordinary glass.

I summarize what we have seen so far: light can be absorbed by the electrons in a material if they can move to quantum states whose energies are higher than the initial energy by an amount equal to the energy of the photon. If there are no such states available in the band structure of a given material, the photons go through the material: it is transparent to light of that colour. This picture also tells us why a material can be transparent to waves in one part of the electromagnetic spectrum and opaque in another part.

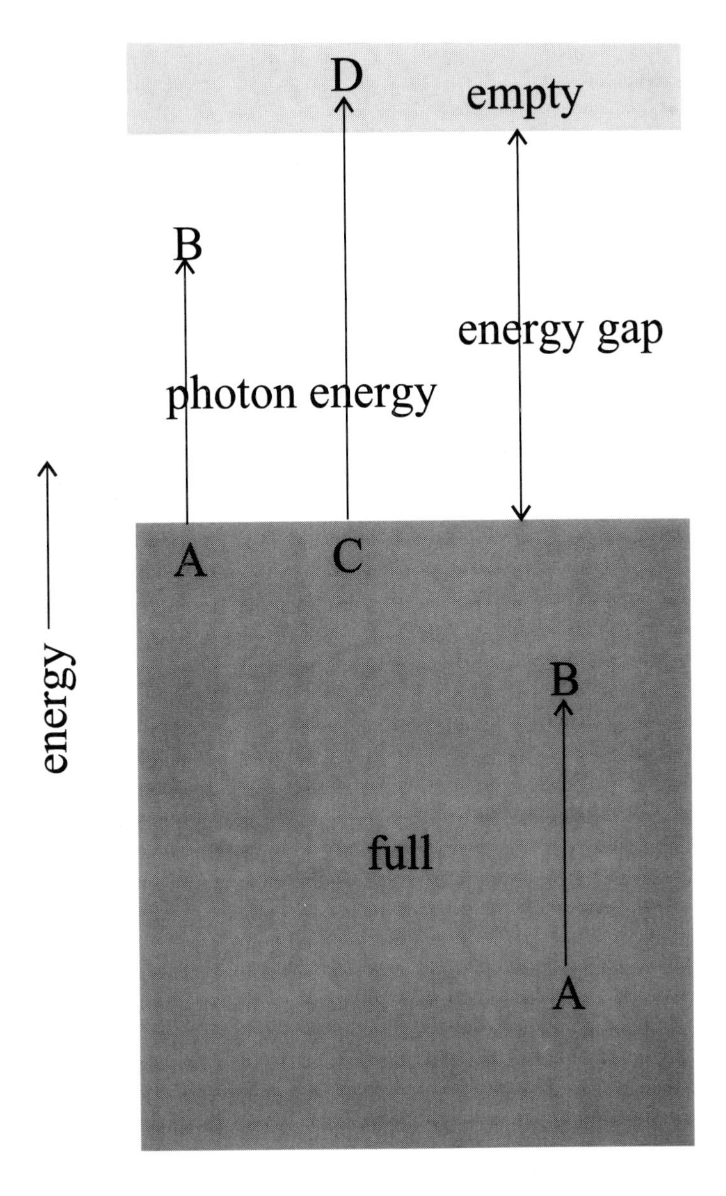

FIGURE IX-7. The electronic energy bands in a transparent material like glass. Photons with energies like AB, that are less than the energy gap between the occupied states and the empty states, cannot be absorbed by the electrons because they have no final states available of the right energy. Such photons will pass through the material, while those with higher energies, like CD, are absorbed.

Now let us look at the case of light falling on the surface of a metal. We know that the metal conducts electricity well, because the outer electrons of each atom are free to wander all through the metal. Figure IX-8 shows a beam of light reflected from the surface of the metal. The free electrons closest to the surface feel a force due to the oscillating electric field in the light wave, and so they start oscillating at the same frequency as the light. One of these electrons, labelled 1, is shown in the figure. This oscillating layer of electrons is taking energy from the light wave, which therefore is weakened and produces smaller oscillations in the next layer of electrons (labelled 2), and so on,

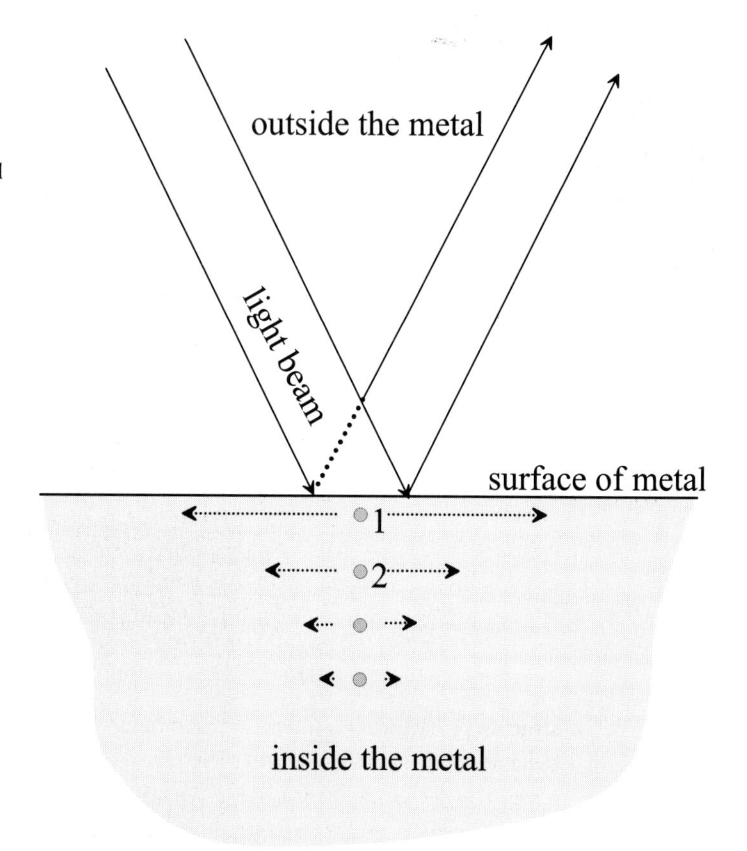

FIGURE IX-8. Absorption and reflection of light for a metal. The oscillating electric field of the light wave sets the conduction electrons (four shown with two labelled 1 and 2) in oscillation. The amplitudes of the oscillations are suggested by the arrows. The electrons thus absorb some of the photons' energy and re-radiate the rest.

until at some depth below the surface the light wave and the electron oscillations have both died out. This depth in most metals is very small: a few tens of angstroms (recall that one angstrom is 10^{-8} cm). So we can see now why even a very thin sheet of metal lets no light through.

You might wonder where the light which is reflected from the surface of the metal comes from, since it looks as if all the light falling on the surface got absorbed in setting up the oscillations of the electrons. You recall that electromagnetic waves are generated by oscillating electric charges. The surface layers of electrons are oscillating at the same frequency as the incoming light. These oscillations produce electromagnetic waves of the same frequency, and these waves are the reflected light.

What I have described so far seems all right for a metal like aluminium, which reflects daylight without changing the colour of the reflected light. But what about metals like copper and gold, which look red and yellow respectively in ordinary daylight which has all the colours of the visible spectrum? The answer lies in a special feature of the electronic band structure of these metals, which I illustrate in fig. IX-9. The conduction band, labelled band 1 in the figure, has all its quantum states occupied by electrons up to the Fermi energy, and the states of higher energy are unoccupied. This part of the band

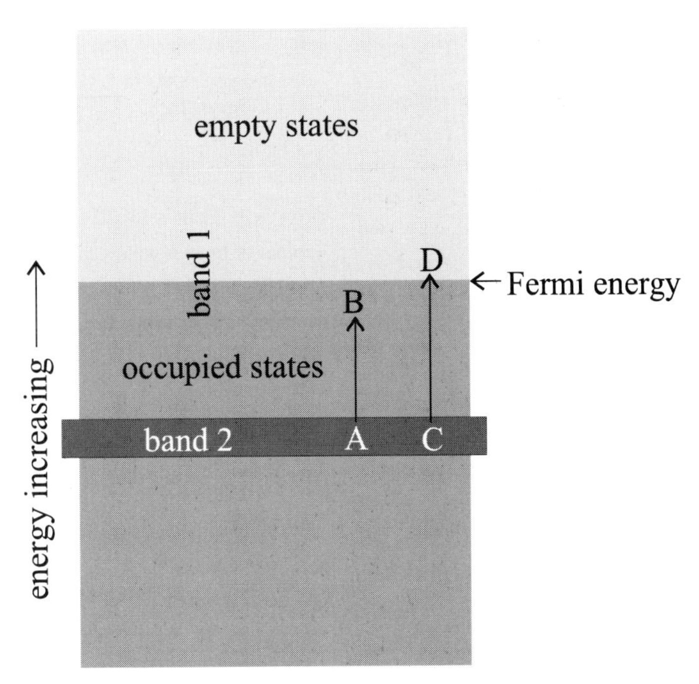

FIGURE IX-9. The electronic energy bands in copper and gold. A narrow energy band, labelled band 2, overlaps the occupied states in the conduction band labelled 1. The large number of electrons in band 2 absorb photons of energies equal to or greater than CD, thus removing these colours from the reflected light. A photon with an energy like AB cannot be absorbed by these electrons. The result is the colour that we see of copper and gold.

structure is the same as that for aluminium, and should therefore produce the same behaviour with respect to light. There is however a new feature in this figure: a second band (labelled band 2) spread over a narrow range of energies which are lower than the Fermi energy and separated from it by a gap in energy. Every quantum state in this band is occupied by an electron, and there are a lot of them, about ten times the number of electrons in the conduction band. Photons with energy of the amount indicated by the line AB in the figure, where B can lie anywhere below the Fermi energy, cannot be absorbed by electrons in band 2, because all the states with this additional energy are already occupied by electrons. These photons can however set the conduction electrons in band 1 into oscillation and eventually lead to reflected light just as with aluminium. On the other hand, photons with energy equal to or greater than the amount corresponding to CD can be swallowed up by electrons in band 2 which then move to previously empty states above the Fermi energy. The reflected light will therefore be missing all these photons and will be different from the light which hit the surface of the metal. The critical photon energy corresponding to CD in fig. IX-9 corresponds to yellow for copper metal, and to green for gold metal. This means that all the colours in daylight from yellow to blue are absorbed in copper, and only red is reflected: hence the red colour of the metal. By similar reasoning, you can see why gold has a yellow colour.

QUESTION The electron which went from the state C to the state D by absorbing the energy of a photon can later go back to C emitting an identical photon. So why should there be any change in the colour of the reflected light?

ANSWER Yes, the electron will go back to C; but the photons emitted in this process come out in all directions, and so the number in the direction of the reflected beam will be much smaller than that in the incoming beam of light.

QUESTION Brass is an alloy made by dissolving some zinc in copper. The colour of brass is yellow instead of red. Why?
ANSWER The addition of zinc to copper changes the energy band structure such that the critical energy CD becomes larger than it was in pure copper, changing the colour corresponding to the energy CD from yellow (as in copper) to green. Thus only the reddish-yellow part of the spectrum is reflected, and the rest is absorbed, by brass.

QUESTION How does one explain the red colour of ruby or the blue of sapphire, which are insulators and therefore have no free electrons?
ANSWER Both ruby and sapphire are a crystalline form of a compound containing two atoms of aluminium for every three atoms of oxygen. In its pure form this compound is called corundum, and is a clear colourless solid. Ruby is a form of this compound containing about one percent of chromium atoms replacing the same number of aluminium atoms, and sapphire has iron atoms instead of chromium atoms replacing the aluminium atoms. To be specific, I shall consider ruby. Each chromium atom is surrounded by aluminium and oxygen atoms, whose influence changes the energy levels of the chromium atom from what they were in the isolated atom. These modified levels are so distributed that most photons with wavelengths shorter than that for red light are absorbed. So when ruby is viewed in ordinary light, only the red comes through. In sapphire, the iron atoms play the same role, except that now the energy levels are different from what they were for chromium, and all portions of the spectrum other than the blue are absorbed by the crystal.

It is the interaction of the photons with the electrons in the material which determines what happens to the light falling on glasses, metals, and minerals. The details depend on how the quantum energy levels of the electrons are arranged in a given material. In a material which is made up of molecules, there are other energy levels due to the internal motions of each molecule. A photon of the right energy would be absorbed by a molecule which thereby moves to one of these higher energy states. To summarize:

If a photon of a certain energy, or equivalently an electromagnetic wave of a certain frequency, is absorbed by a given material, then one can conclude that the absorbed energy has put the material in a new quantum state whose energy is higher by exactly that amount. There can be no absorption of the photon if such a quantum state with the right energy is not available.

The most widely occuring molecule in the world is the molecule of water, consisting of two atoms of hydrogen joined to one atom of oxygen. The atoms vibrate about their average positions, and the molecule as a whole rotates about an axis. These motions lead to a spectrum of quantized energy levels which are in addition to the electronic energy levels of each atom in the molecule. You can imagine that all this leads to a complex set of energy levels capable of absorbing photons of a wide range of frequencies. There is one remarkable exception which makes life as we know it on this planet possible. In that multitude of energy levels, there are very few in that narrow range which would cause photons of visible light to be absorbed. Figure IX-10 shows the thickness of water needed to absorb 99 per cent of the electromagnetic waves of different frequencies passing through it. Over most of the range of frequencies from microwaves to X-rays, only a fraction of a centimetre of water is enough to absorb almost all the radiation. For the narrow range of frequencies of visible light, however, the absorption becomes very weak: it takes tens of metres of water to absorb 99 per cent of such light. When water is in the form of vapour, as in the atmosphere, there are fewer molecules per metre of thickness, and the absorption is therefore even less. If this narrow window of transparency of water to visible sunlight did not exist, such light would not have reached the surface of the earth, and no life as we know it could have evolved. Given that it is solar energy which sustains life on earth, one might say that a quantum-mechanical peculiarity of the energy levels in water permitted that particular form of life to evolve on earth which utilizes the photons of visible light. One might speculate on the form that life might have taken if the window of transparency lay in some other part of the solar spectrum: such is the stuff of science fiction.

The microwave oven found in many kitchens is an application of the absorption of electromagnetic waves by water and fat. The oven contains a generator which produces standing microwaves at a frequency of about 2×10^9 hertz in the chamber where the food is placed. These photons are absorbed mostly by the molecules of water and fat

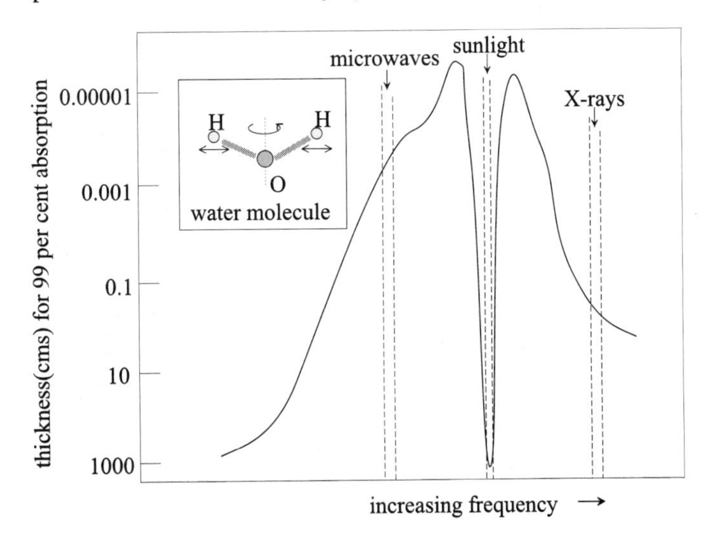

FIGURE IX-10. The thickness of a sheet of water which will absorb 99 per cent of electromagnetic waves passing through it, at different frequencies. The inset shows the kinds of rotational and vibrational motions (indicated by arrows) of a water molecule, whose quantum energy levels are responsible for this absorption.

that are present in all foodstuffs, thus raising their energy and temperature. The plastic or glass of which the container is made does not have those energy levels which would permit the absorption of these photons, and therefore remains relatively cool. The conduction electrons in a metal, on the other hand, can absorb these photons and thus heat the metal, as anyone who has mistakenly left a metal spoon in the oven will have noticed.

QUESTION Why must the food be placed on a rotating platform in a microwave oven, but not so in an ordinary oven?

ANSWER Since there are standing waves in the microwave oven, with a wavelength comparable to the size of the oven, the strength of the electromagnetic field varies from zero to its maximum value at different points inside the oven. Without rotation, the food would be heated non-uniformly. The heating in an ordinary oven is due to infrared radiation of much shorter wavelength, and the problem of non-uniform heating does not arise.

6 Refraction of light

We have seen in the previous section how the existence of quantum energy levels in a solid leads to the absorption of electromagnetic waves as they pass through it. Such absorption means that the radiation progressively loses energy as it moves through the solid. The radiation has some other characteristics too: it has a wavelength, a frequency and a speed. Fixing any two of these quantities automatically fixes the third, because frequency multiplied by wavelength equals speed of wave. So let us see what happens to the frequency and the wavelength when radiation passes from empty space into a transparent material like glass, where the absorption is negligible.

The frequency of the radiation is determined initially by the source: the sun, or an electric light, for example. When this radiation passes through the material, the oscillating electromagnetic field acts in opposite directions on the charged electrons and nucleus of each atom and pulls them apart a bit. The atom in this condition is said to be in a state of *polarization*, or is *polarized.* Because the field which produces the polarization is oscillating in magnitude as well as direction, so will the polarization itself, and at the same frequency too. This oscillating polarization produces radiation at the same frequency, and so the process continues: a complicated interaction between the electromagnetic field of the radiation and the charged electrons and nuclei in the material. In any event, the conclusion is that the frequency is unchanged in the material.

How about the wavelength of the radiation when it is in the material? Figure IX-11 shows an electromagnetic wave travelling from air into a crystal. For the present purpose, air can be treated as empty space, because the wavelength is practically the same in both, and we denote it by $\lambda(0)$, as indicated in the left half of the figure. If the

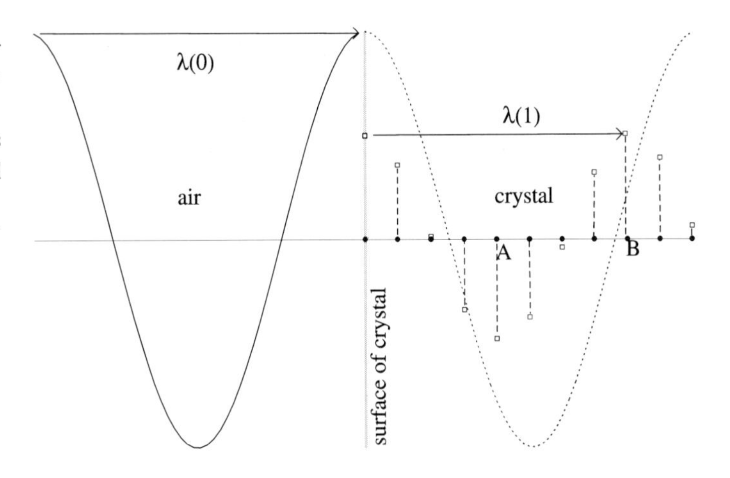

FIGURE IX-11. A light wave of wavelength λ(0) in air enters a transparent crystal, in which its wavelength is reduced to λ(1). The atoms, shown as black dots like A and B, are polarized to different extents by the electric field of the light wave, which is itself modified by the polarization. The amount of polarization of each atom is suggested by the distance and direction of the hollow square from the black dot representing the atom. The result is a shortening of the wavelength in the crystal.

wave had gone on through the crystal without being affected by it, then it would be like the broken line in the right half of the figure, with the same wavelength λ(0). But in reality the wave is affected by the crystal, because it produces displacements of the electrons and nuclei which make up the atoms. The magnitude and direction of this atomic polarization varies from atom to atom, and I have indicated this by a small square against each atomic position at a distance proportional to the polarization. For example, the polarization of atom A is about equal in magnitude but opposite in direction to that of atom B. The electric field is tied to the polarization at each atom, and so we have an oscillating electric field with the same wavelength as the oscillating polarization, and this is then the light wave in the crystal. Note that the light wave in the crystal is associated with the movement of things with mass, namely electrons and nuclei. This is in contrast to the same light wave in free space, where there is no mass to be moved. It is this difference which leads to the wavelength of light being shorter in matter than in free space. Since speed is equal to frequency multiplied by wavelength, the speed of light in any material medium is less than its speed in free space.

The fact that the speed of light is less in a solid than it is in free space causes a beam of light to change its direction as it goes from one to the other. This effect, known as the *refraction* of light, is what makes possible a host of optical equipment: eyeglasses, cameras, telescopes, and microscopes, to name a few. I illustrate how refraction comes about in fig. IX-12. A beam of light comes from the left and enters a transparent solid. If there were no solid, the wave would continue in the same direction, as shown by the broken lines. The motion of the wave is perpendicular to the wavefront, as you can see for example with waves on water. Now consider the wavefront AB in the figure. The part of the wave at B has just reached the surface of the solid, while the part A still has the distance AC to go before it reaches the surface. During the time it takes for this to happen, the part of the wave at B will have travelled to D, and the distance BD will be less than the distance AC because of the lower speed in the solid. The wavefront in the solid will be the line CD, and the wave will be travelling at right angles to this line.

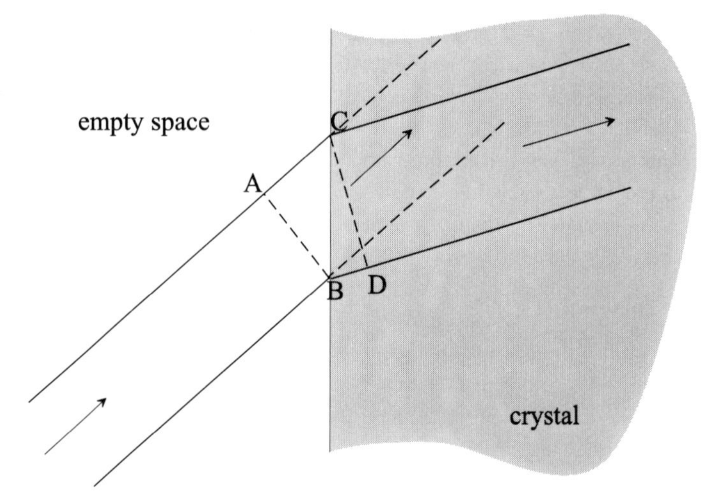

FIGURE IX-12. Refraction of a beam of light on entering a crystal from empty space. The light takes the same time to get from A to C as it does from B to D, because its speed in the crystal is less than its speed in empty space.

This last condition can be satisfied only if the beam in the solid is in a specific direction, and none other, relative to the beam in empty space.

QUESTION In terms of the lengths in fig. IX-12, what is the speed of light in the solid, v, in terms of the speed in empty space c?

ANSWER Light travels the distance AC in free space in the same time that it travels the distance BD in the solid. Therefore the ratio of v to c is equal to the ratio of the length BD to the length AC.

QUESTION Is it possible to see from fig. IX-12 why a pool of water appears shallower than it actually is when viewed from above?

ANSWER Yes. Imagine the solid replaced by water, and the arrows showing the direction of the beam of light to be reversed. So now the light is coming from the bottom of the pool, say, and its direction is changed as shown when it emerges from the water. A person looking into the water will therefore see the bottom in the direction shown by the broken lines. The bottom will thus appear to be nearer to the surface of the water than it actually is. In fact, the actual bending of the beam of light as it emerges from water to air is such that a pool which is two metres deep will seem to be only a metre and a half deep: inexperienced swimmers, beware!

7 Dispersion of light

The speed of an electromagnetic wave in empty space, whatever its frequency, is the same, and we denote it by the symbol c, with a value of 3×10^{10} cm per second. When the wave travels through a material medium, not only is its speed reduced, but the amount of reduction depends on both the frequency and the medium. This effect,

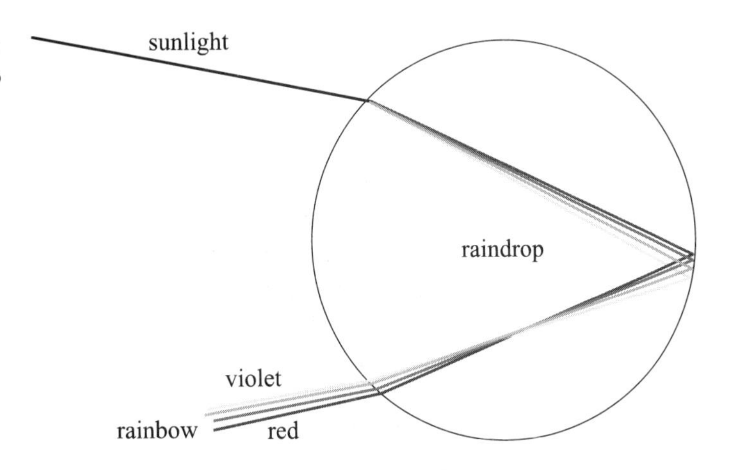

FIGURE IX-13. How a ray of sunlight is refracted and reflected in a raindrop to produce a rainbow. The slight differences in the speed of light in water for different colours is an essential factor in this phenomenon.

called *dispersion,* is greater the more densely packed are the atoms in the medium. For example, the dispersion of light in air is negligibly small, so that light of any colour has practically the same speed, c, in air. In water or in glass, on the other hand, there is a significant decrease in speed which is different in the two substances and for different colours. The amount of bending of light as it enters (or leaves) water or glass from air therefore depends on the colour. This is the origin of the rainbow, and the spectrum of colours produced when sunlight passes through a glass prism, to name two examples of dispersion. I illustrate these effects in figs. IX-13 and IX-14.

Figure IX-13 shows the path of a ray of sunlight as it enters a raindrop, is reflected once and emerges into the air again. The parts of the ray corresponding to different colours are bent differently during this passage, and emerge in slightly different directions as shown. If I stop at this point, I would have no rainbow, for the following reason. Suppose the ray of sunlight entering the raindrop at a particular spot is sending red light in my direction. A ray entering at another spot may send blue light in my direction, and so on, so that in the end my eye receives all the colours and therefore sees just white light.

But this does not happen. The rules which govern reflection and refraction of light combine with the geometrical properties of the sphere (which is the shape of the raindrop) to make the light coming out in a particular direction (which changes slightly with colour) to be more bright than in all other directions. The result is the rainbow. The fact that the speed in water of blue light is lower than that of red light is what ultimately causes blue to be on the inside and red on the outside of the rainbow. Only a part of the light emerges from the raindrop after the first reflection. The rest is reflected once more at the raindrop's surface, and a part of this emerges and gives rise to the second rainbow which can sometimes be seen. The order of colours in this rainbow is the reverse of that in the first. This fact of nature has been overlooked by some artists famous for their landscape paintings. The Neue Pinakothek in Munich for example has two landscapes with double rainbows, painted by Jan Wildens and by Josef Anton Koch, where the colours are not in the order that nature says they should be.

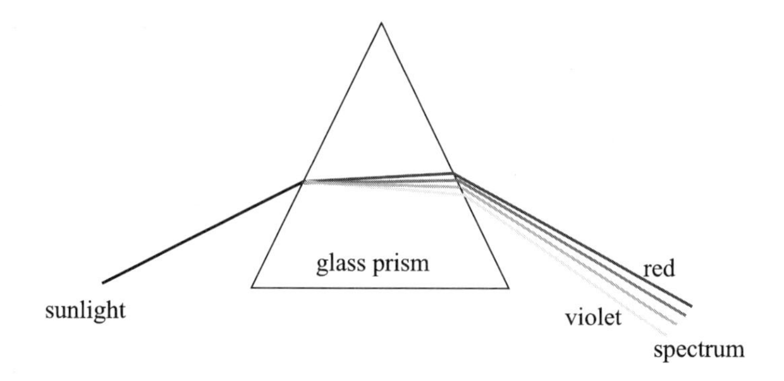

FIGURE IX-14. How a glass prism produces the spectrum of sunlight through two successive refractions.

Figure IX-14 shows how a triangular glass prism breaks up sunlight into its constituent colours. This is the basic effect which is responsible for the sparkling colours of cut diamonds and the beads in a glass chandelier, for example. It is used in the construction of one type of an instrument called a *spectroscope*, which shows the spectrum of frequencies in light which passes through it. We have seen that each type of atom – oxygen, carbon, etc. – has its own characteristic set of energy levels which is different from that of all other types. When the atom absorbs some energy (for example when the material is heated sufficiently), it goes to a higher energy level and subsequently returns to a state of lower energy. The difference in energy E is given off by the atom as a photon, which is light of a certain frequency f related to this energy through the Planck constant h:

$$E = hf.$$

Thus the light coming out has a spectrum of colours with certain frequencies which are characteristic of the atoms producing it. Knowing the spectrum, one can deduce which type of atom it comes from. The rainbow is the spectrum of the thermal radiation from the sun produced by a spectroscope that occurs in nature, namely rain.

8 Solar batteries and LEDs

A solar battery is a semiconductor device which converts some of the energy it receives from sunlight into electrical energy, just as an ordinary battery converts chemical energy into electrical energy. The LED (acronym for *l*ight-*e*mitting *d*iode) is a semiconductor device which does the reverse: it uses electrical energy to produce light. A solar battery is a practical application of what we have learnt about the energy bands in semiconductors.

Figure IX-15 shows how a solar battery works. In part (*a*), an n-type and a p-type semiconductor are close together but not in contact. The n-type has electrons, and the p-type has holes, moving about. In part (*b*), the two semiconductors are brought into contact to form what is called a *p-n junction.* The electrons and holes at the junction

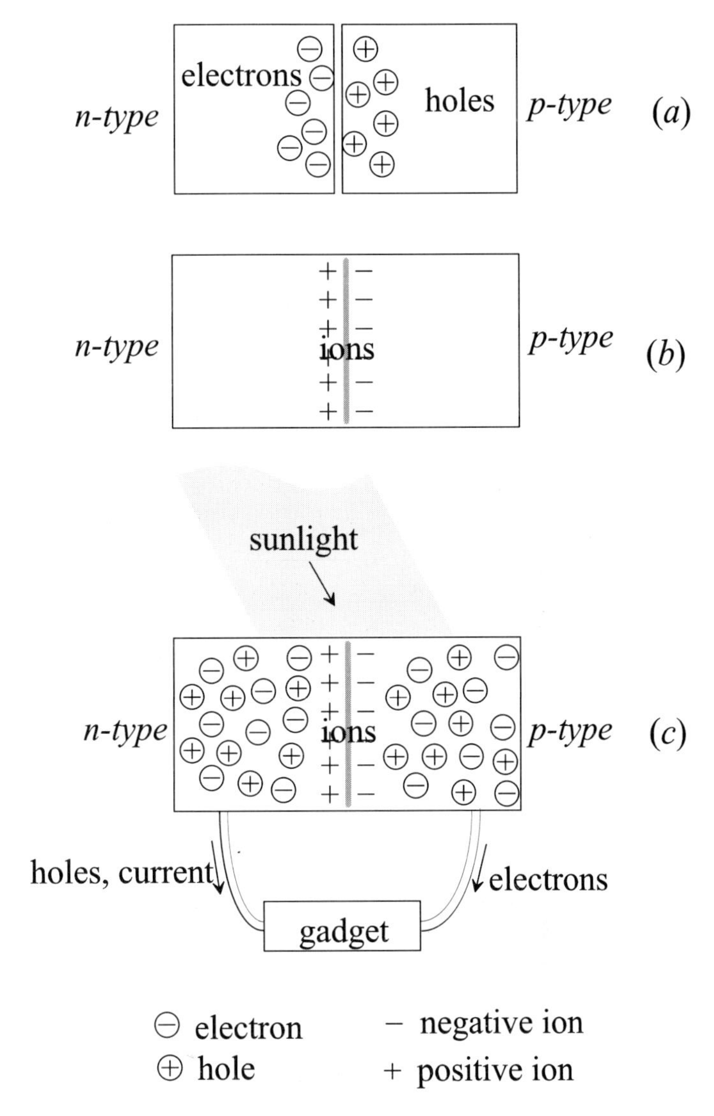

FIGURE IX-15. The working of a solar battery. (*a*) Isolated p- and n-type semiconductors. (*b*) A p-n junction. The electrons and holes neutralise each other at the junction, leaving a layer of positive ions and one of negative ions facing each other and forming, as it were, the positive and negative terminals of a battery. (*c*) Sunlight falling on the junction creates additional electrons and holes which are driven as an electric current through a gadget connected to the battery with wires.

move across and neutralize each other, leaving positive ions on one side and negative ions on the other.

QUESTION Why does the process not continue, thus neutralizing all the moving charges on both sides?

ANSWER An electron from the left which tries to move through the junction meets the phalanx of negative ions on the right and is repelled; it cannot get through. The positive holes on the right are similarly stopped by the positive ions on the left. Remember that like charges repel one another.

The situation in part (*b*) is what obtains in a solar battery. When sunlight falls on this device, as in part (*c*) of the figure, some of the photons give electrons in the valence band enough energy to move them to the conduction band, leaving empty states in the valence band. The electrons in the p-type and the holes in the n-type feel a force due to the ions at the junction shown in part (*b*), which makes them flow in the directions shown through a gadget (which could be a light bulb or motor or whatever) connected with wires to the device. The direction of the electric current itself is taken to be opposite to the direction of electron flow and in the direction of the hole flow, and is therefore as shown in the figure.

QUESTION What happens to the electrons in the n-type and the holes in the p-type which are left behind?

ANSWER: They find their way to the region of the junction, where they neutralize each other.

9 Summary

This chapter deals with the nature of light, and how it interacts with material bodies. In section 2, we are reminded that light manifests itself as either waves or particles, depending on the mode of observation: it has the property of duality, just like electrons, protons, neutrons, etc. Visible light is just a small part of electromagnetic radiation, whose wavelengths can range from 10^{10} cm to 10^{-14} cm and even beyond these limits. All these waves travel at the same speed through empty space of about 3×10^{10} cm per second, denoted by the symbol *c*. This speed is one of the constants of nature.

I introduce the idea of a *field* in section 3 with the examples of the temperature and the velocity of circulation of air at different points in an oven. I then describe electric and magnetic fields, and show how light waves (as well as all other electromagnetic waves) consist of these fields. In what follows, I consider mostly visible light, but the basic ideas apply with suitable modifications to all electromagnetic radiation.

I describe in section 4 two ways in which light can be generated. In one, the quantum state of an atom or molecule changes to one of lower energy, and the difference in energy is carried away by a photon. Since the energy levels are discrete, the resulting photons can have only certain frequencies and nothing in between. An example is the light from a neon tube. The other way to get light is when the energy of motion of the constituent atoms, electrons, etc. due to the temperature of the material is transformed into photons. The energies of these motions are quantized, but lie so close together that the resulting photons have a practically continuous distribution of frequencies, examples of which are the light from the sun or from an incandescent electric lamp.

Some materials are transparent to light, and others are opaque. I show in section 5 how this difference depends on the relationship between the photon energies and the quantum energy levels in the material. This relationship also explains the colours of materials like copper, gold, and minerals.

The speed of light depends on the substance through which it is passing, and is highest in empty space. A consequence of this is *refraction*, which is the change in direction of a light beam in passing from one substance to another. How this comes about is described in section 6.

In section 7, I find that the reduction in the speed of light is different for different wavelengths and substances. This is called the *dispersion* of light. This is the origin of the rainbow, and the principle is used in an instrument called a spectroscope which analyses light from a source into its constituent wavelengths.

I describe in section 8 a practical application of the interaction of light with matter, namely a solar battery. We see how the nature of the quantum energy levels in a semiconductor p-n junction helps to convert the energy of the photons in sunlight into electrical energy.

This concludes my description of electromagnetic waves and their interaction with materials. It is also possible to have electric and magnetic fields which are not part of an electromagnetic wave: an electric battery, for example, is the source of an electric field, and a magnet produces a magnetic field. Such fields also produce effects in a piece of matter, because it has electrons and nuclei which not only have electric charge but are also like permanent magnets. I shall look at some of these effects in the next two chapters.

X ELECTRIC BULBS AND INSULATED CABLES

1 Preliminaries

Electricity is one of the most widely used forms of energy: in the home, in industry, for transportation and for communication, to name just a few of its applications. It is generated at one place and transported to another which may be hundreds of kilometres away. It is carried in cables made of copper or aluminium strung all over the landscape. Electricity inside a home is at about 230 volts in most places and at 115 volts in North America. It is distributed by copper wires which are covered with some sort of insulation. The rate at which electrical energy is used in appliances is measured in watts: a 100-watt bulb, a 2000-watt furnace, and so on. I show all this, stripped down to the barest essentials, in fig. X-1. The generator produces electricity, which is carried through cables and wiring and finally arrives at the appliance where it is used. The figure also suggests an analogy to a hydraulic system. Table X-1 sets forth this analogy: similar items are set side by side in the two columns.

In the hydraulic (electric) system, the pump (generator) raises the pressure (voltage) of water (electrons), consequently raising its (their) energy. The stream of water (electrons) flows down the pipe (cable) to the turbine (toaster), where the energy is converted to kinetic (thermal) energy. The water (electrons) returns (return) to where it all started but with lower energy, and the whole cycle continues. Notice that there is no loss of the working substance in either case. It just keeps going round and round the circuit. There is one difference between the two systems, which is not important for the analogy I have drawn. The electric current is oscillating back and forth 50 times (60 times in North America) per second, whereas the water keeps going in the same direction all the time.

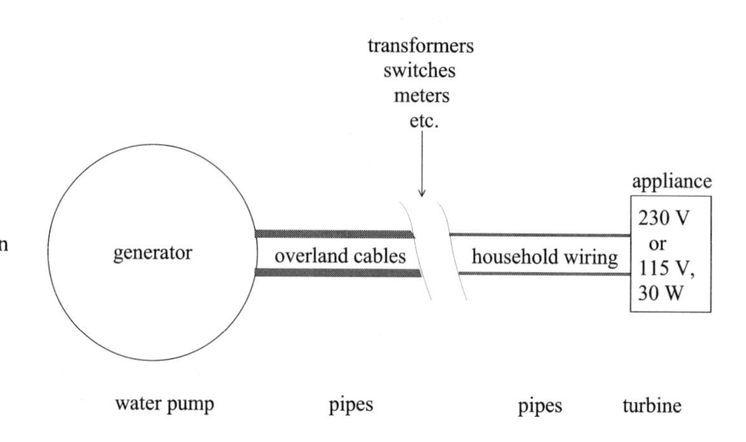

FIGURE X-1. A schematic picture of an electricity supply. Only the generator and the household appliance are shown; practically everything in between is in the open gap. Underneath are named the corresponding parts of a hydraulic system which uses water under pressure to drive a turbine. The analogy is explained in the text.

Table X-1. *Analogy between electric and hydraulic systems*

hydraulic system	electric system
working substance: water	working substance: electrons
water pump	electric generator
space inside pipe which carries the water	metal (copper, aluminium) wire which carries the electricity
water meters	electricity meters
pressure	voltage
appliance, e.g. turbine, which converts potential energy of water into kinetic energy of turbine	appliance, e.g. toaster which converts electrical energy into thermal energy
water current (amount of water per sec)	electric current (amount of electric charge per sec)

I shall say a few words about volts and amperes, which are respectively the units for measuring voltage and electric current. If the voltage of a battery is two volts, for example, it means that the difference in potential energy of an electron in moving from one terminal of the battery to the other is the charge multiplied by the voltage, or $2e$, where e is the magnitude of the electron's charge. It is this energy which gets converted to other forms of energy when the electron goes through the appliance. The electric field and the voltage are closely related. In a battery, for example, there is a voltage between the terminals, and there is an electric field everywhere in the space between the terminals.

The electric current is the amount of charge that is circulating per second past any part of the circuit, and we have just seen that it is the same everywhere in the circuit: in the wires connected to the wall socket as well as through the toaster, for example. It is measured in amperes. One ampere is a current of about 6×10^{18} electrons per second.

In this chapter, we look at what happens when an electric field (or equivalently a voltage) is applied to a solid, some of whose electrons are free to move about under the influence of the field and thus produce an electric current. This means that we shall be concerned mostly with metals and semiconductors, since insulators do not have mobile electrons. We shall find out how the electronic band structure of metals and semiconductors leads to their characteristic electrical properties, and why it is more difficult to pass an electric current through an alloy than through a pure metal.

2 Electrical resistance

I consider a thin wire of tungsten connected to a battery. The electric field between the terminals of the battery exerts a force on the electrons in the metal, causing them continually to pick up speed in the direction of the positive terminal. If there were nothing to prevent it, the speed would continue to increase, and so would the current. But with a long and thin enough wire, the current quickly reaches a constant value and stays there. I find that the size I of this current is proportional to the voltage V. So the voltage divided by the current is a constant:

$$\frac{V}{I} = R.$$

The quantity R is called the *electrical resistance* (or often simply *resistance*) of the wire, and is in *ohms* if V is in volts and I in amperes.

The resistance of a wire depends on its size and shape. A meaningful comparison of the resistances of different materials is therefore possible only if they all were wires of identical size and shape. Scientists have agreed on what this shall be: a cube of the material with an edge of one cm. Its resistance measured between a pair of opposite faces is called the *resistivity* of the material, and is expressed in *ohm-cm.*

QUESTION An electric bulb carries a current of half an ampere at 220 volts. What is its resistance?

ANSWER The voltage is equal to the resistance multiplied by the current. Therefore the resistance is the voltage divided by the current, or 220 divided by a half, or 440 ohms.

We now want to understand how the quantum description of a metal and a semiconductor explains the features of their electrical resistance, and why an insulator does not carry a current when a voltage is applied to it. I describe in section 3 the essential facts, and go on to their explanation in the succeeding sections.

3 The observed properties

The first property to note is that the resistivity is different for different materials. Table X-2 shows the resistivities of some metals, a semiconductor, and an insulator, all measured at 20 °C. You can see how greatly the three types differ from one another in their

Table X-2. Resistivities in ohm-cm of various materials at 20 °C

material	resistivity in ohm-cm
copper (metal)	1.7×10^{-6}
silver (metal)	1.6×10^{-6}
lead (metal)	22×10^{-6}
germanium (semiconductor)	1 (value is very sensitive to impurity content)
glass (insulator)	10^{12} (depends on type, but still very large)

resistivities. There is no other property of a solid which covers such a vast range of values, differing at the extremes by a factor 10^{18} (or even more if low temperatures and superconductors are included).

The effect of a change of temperature on the resistivity of metals and semiconductors is shown in fig. X-2. The resistivity of a metal drops with decreasing temperature, appearing to head towards zero at the absolute zero of temperature. However, this decrease levels off at a low temperature of a few kelvin. In a semiconductor like germanium, the effect of lowering the temperature is the opposite: the resistivity increases. There are, in addition, many metals and alloys whose resistivity, instead of levelling off, drops abruptly to zero at some temperature which is different for different materials. These are called *superconductors*, and are the subject of chapter XII.

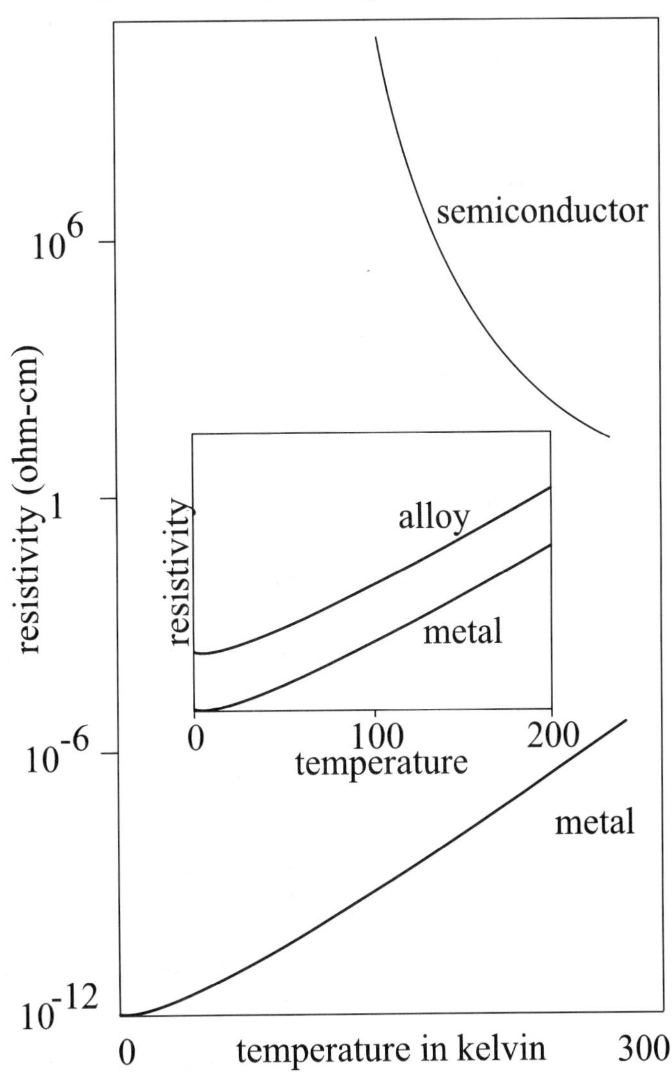

FIGURE X-2. The variation of electrical resistivity with temperature for a pure metal, an alloy, and a pure semiconductor. Note the logarithmic scale for resistivity in the larger picture which compares a metal and a semiconductor. The inset compares the resistivities of a pure metal and an alloy based on the same metal, and the resistivity scale here is the usual linear one. The resistivity of the alloy is greater than that of the metal by a constant amount at all temperatures.

I look next at the effect of adding small amounts of a second element to a metal and a semiconductor. Very roughly, one can say that adding one per cent of a second type of atom to a pure metal increases its resistivity at all temperatures by a constant amount of about one microhm-cm, which is 10^{-6} ohm-cm. The inset in fig. X-2 shows the typical behaviour of a metallic alloy. The effect of an added second element on a semiconductor is more dramatic and complicated. Its resistivity can drop by a factor of as much as a million, and the drop can be different at different temperatures.

4 Scattering of electrons

Suppose that I think of the electrons as particles moving about in a metal wire and bouncing off the ions in the lattice. When I apply a voltage to the wire, the electrons accelerate down the wire, increasing their energy and producing a current. They would still be colliding with the ions and thus losing some energy, and eventually some steady current should result. I might think that this is the way electrical resistance appears. It is like driving a car through a forest without roads. Progress would be somewhat slow because of collisions with the trees.

Nothing about this picture should change if I cool the wire to a lower temperature, and I would therefore expect the resistance to be unchanged. But, as fig. X-2 shows, this is not what happens: the resistance drops dramatically as the temperature drops. The resistance of a pure metal at a few kelvin may be a million times smaller than that at 300 kelvin. It is quite clear that there is something wrong with the picture of particle-like electrons bouncing off ions to give rise to electrical resistance. What is wrong is that I have not used quantum mechanics to describe the electrons. I have not taken account of their wave-like nature. A better analogy than the car in the forest would be a person trying to call out to someone else in the forest; the trees do not greatly impede the sound waves.

We saw in chapter VI that electrons in crystalline solids are represented by Bloch waves. As shown in fig. VI-10, it is a wave, with some additional wiggles at the position of each ion. It is the appearance of these wiggles, and not scattering, that is the result of the interaction between the charge of the electron and charges of the ions. Once we take Bloch waves to represent the electrons, we can forget about the ions. The Bloch waves can be thought of as being in empty space, and there is nothing to hinder the motion of the electrons they represent. Then what else could it be which scatters the electrons, thus preventing a runaway current when a voltage is applied?

I said that we could forget about the ions. Well, not quite: it is only the static ions that we can forget. There are still the vibrations of the ions about their mean positions. It is the kinetic energy of these vibrations which constitutes the thermal energy of the lattice and gives the solid its temperature. We saw in chapter VI that the quantization of these vibrations gives us phonons. So the Bloch waves are not in empty space, but in a space filled with these particles called phonons, all moving about at the speed of sound in the solid. It is the scattering of the electrons when they collide with the phonons which limits the electric current, and leads to a resistance.

How does the electron know that there is a phonon? A phonon is a quantum of a sound wave, which is a wave of alternating compression and expansion of the lattice of ions. The conduction electrons keep in step with these oscillations of the ionic lattice, but not perfectly. The result is that the compressed (expanded) parts of the sound wave have a net positive (negative) charge. The sound wave is thus accompanied by a superposed wave of charge oscillating between positive and negative values. This charge wave influences the conduction electrons and causes them to change from one quantum state to another. This is the mechanism of scattering of electrons by phonons.

Electrons are fermions, and this means that each quantum state with a specified **k** vector can have no more than two electrons with opposite spins. I described in chapter VI (p.118) how this leads to the picture of a Fermi surface inside which lie the wave vectors of these electrons. Remembering that the momentum is equal to the wave vector multiplied by the Planck constant, and also to the velocity multiplied by the mass, I can think of the Fermi surface as giving me a picture of the velocities of the electrons. Figure X-3 shows what happens to a Fermi surface, which I have assumed to be spherical so as not

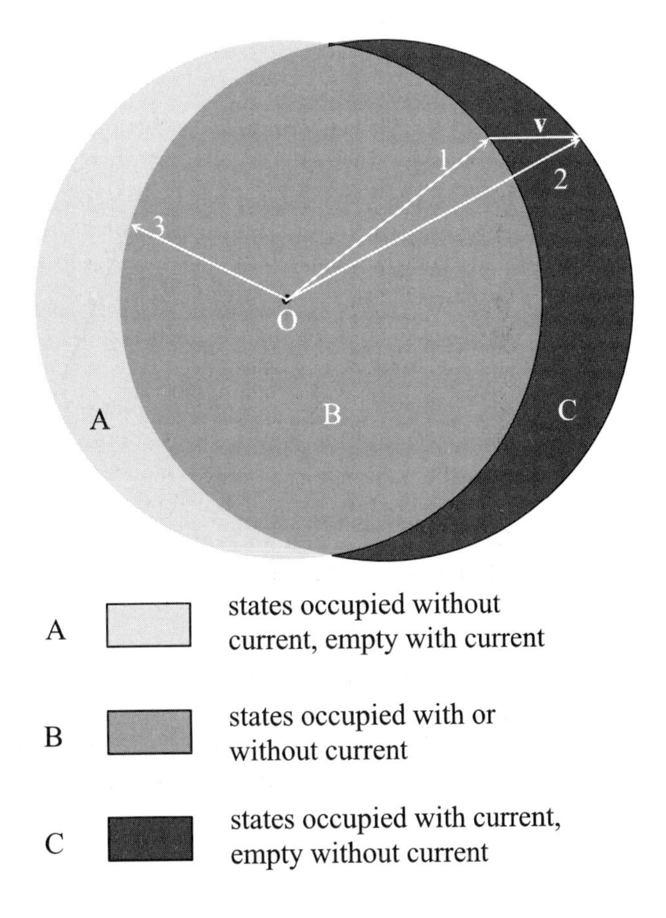

FIGURE X-3. The effect of an electric current on the Fermi surface. The parts A and B together represent the Fermi surface when there is no electric current. In the presence of a current, the Fermi surface is bodily displaced by the vector **v** and is now represented by the parts B and C taken together.

to complicate the picture, when a current flows. What follows applies to any other shape of Fermi surface also. The parts labelled A and B in fig. X-3 taken together form a sphere with its centre at the point O. The velocity vectors for all the electrons lie within this sphere, which shows the state of affairs when there is no current flowing in the metal. Although the electrons are moving about in all possible directions with velocities from zero up to a maximum, there is no net current. For each electron moving in a given direction with a given speed, there is another one moving in exactly the opposite direction with the same speed, and the two add up to give zero current.

Now I apply a voltage and thus exert a force on each electron, causing it to get an added velocity in the direction of the force. It is scattered after a certain time which I denote by the Greek letter τ (tau), at which instant each electron has had its velocity increased by an amount $\mathbf{v}$. The electron whose initial velocity is shown by the vector labelled 1 in the figure, for example, now has the velocity vector labelled 2. At this instant the electron is scattered by a phonon and changes its wave vector to the one labelled 3, say. If you imagine this process occurring with each of the velocity vectors, you can see that the net effect is to displace the whole sphere to the right by the vector $\mathbf{v}$, so that it is now comprised of the portions labelled B and C. But it is not a static picture: application of a voltage moves the sphere a certain amount and no further, because electrons are constantly disappearing, because of scattering, from quantum states at the leading edge in region C and reappearing in states at the trailing edge in region B.

The current is the charge passing a given point per second, and is therefore proportional to $\mathbf{v}$, which is called the drift velocity. The time τ which elapses before the electron is scattered is called the *scattering time.* A longer scattering time means a larger drift velocity and therefore a larger current too. Furthermore, the current is also proportional to the number per unit volume (called the *density*) of electrons which are there to carry it. So we should look to the scattering time and the density of electrons for an explanation of the different behaviours of the electrical resistance of metals, alloys and semiconductors shown in fig. X-2.

We take the case of a pure metal first. When its temperature goes up, so does the amount of thermal energy in it. This energy consists of the vibrations of its atoms about their average positions, and this motion is quantized into phonons. The number of phonons increases with an increase in temperature, and so the frequency of collisions between electrons and phonons also increases. This means that the scattering time (which is the time between collisions) and therefore the conductivity decrease as the temperature rises. In other words, the electrical resistance should decrease with decreasing temperature in a metal, and this is what happens. The electron density may vary from metal to metal, but does not change with temperature in a given metal. We thus have a basic explanation of the different resistivities of different metals, and how they change with temperature.

In an alloy, we have more than one kind of atom in the lattice: copper and zinc atoms in brass, for example. We saw in chapter VIII (p.155) that this leads to an additional scattering of the electrons, due to the loss of translational symmetry. This in turn causes an electrical resistance which is added to that due to the scattering of electrons by phonons. Furthermore, this resistance will not change with temperature, because the two types of atoms do not change their positions. This is the behaviour of the resistance of an alloy that is actually observed, as shown in fig. X-2.

Finally we come to the electrical resistance of a pure semiconductor like silicon or germanium. There are two striking differences from the resistance of metals: it is very large, and gets even larger as the temperature drops. We again look to the two quantities, the scattering time and the number of electrons per unit volume, for an explanation. The scattering time is mainly determined by the number of phonons, which in turn is determined by the temperature of the material. We would thus not expect spectacular differences in scattering times between metals and semiconductors. The situation however is quite different with the number of electrons available to carry the current.

A semiconductor is essentially an insulator: the energy levels for the electrons are distributed as shown in fig. X-4. There is a band of levels, called the *valence band.* Above this band is a range of energy, called the *energy gap*: there are no quantum states for electrons in this gap. Above the gap there is again a band of possible energies, called the *conduction band.* At very low temperatures, almost every quantum state in the valence band is occupied by an electron, and the conduction band has very few electrons. An applied voltage produces negligible current, and thus we have an insulator.

When the temperature is raised, some electrons in the valence band acquire enough additional energy in the form of thermal energy to put them into previously unoccupied states in the conduction band. This additional energy must be equal at least to the energy gap, because there are no states for the electrons in the energy gap itself. An equal number of empty states is left behind in the valence band, which becomes equivalent to a band of holes. The number of electrons and holes increases with increasing temperature. This means that a given applied voltage will produce more current the higher the temperature, because there are more electrons and holes to carry it. Therefore the resistivity of a semiconductor should decrease as the temperature goes up. This is what happens, as can be seen in fig. X-2.

How about the scattering time, which also influences the resistivity of the semiconductor? The scattering is still by phonons, and is stronger at higher temperatures just as with metals. But its effect in changing the resistivity is weak in comparison with the effect of the change in number of electrons and holes, and the net result is a drop in resistivity with increasing temperature, as is actually observed.

The addition of small amounts of other elements to a semiconductor – phosphorus or aluminium to silicon, for example – will result in electrons in the conduction band or empty states in the valence band (equivalent to a band of holes) being present even at the lowest temperatures, as explained in chapter VI (page 121). There will also be thermal excitation of electrons and holes as in a pure semiconductor, and so the variation of resistivity of such doped semiconductors will be somewhere between that of metals and pure semiconductors.

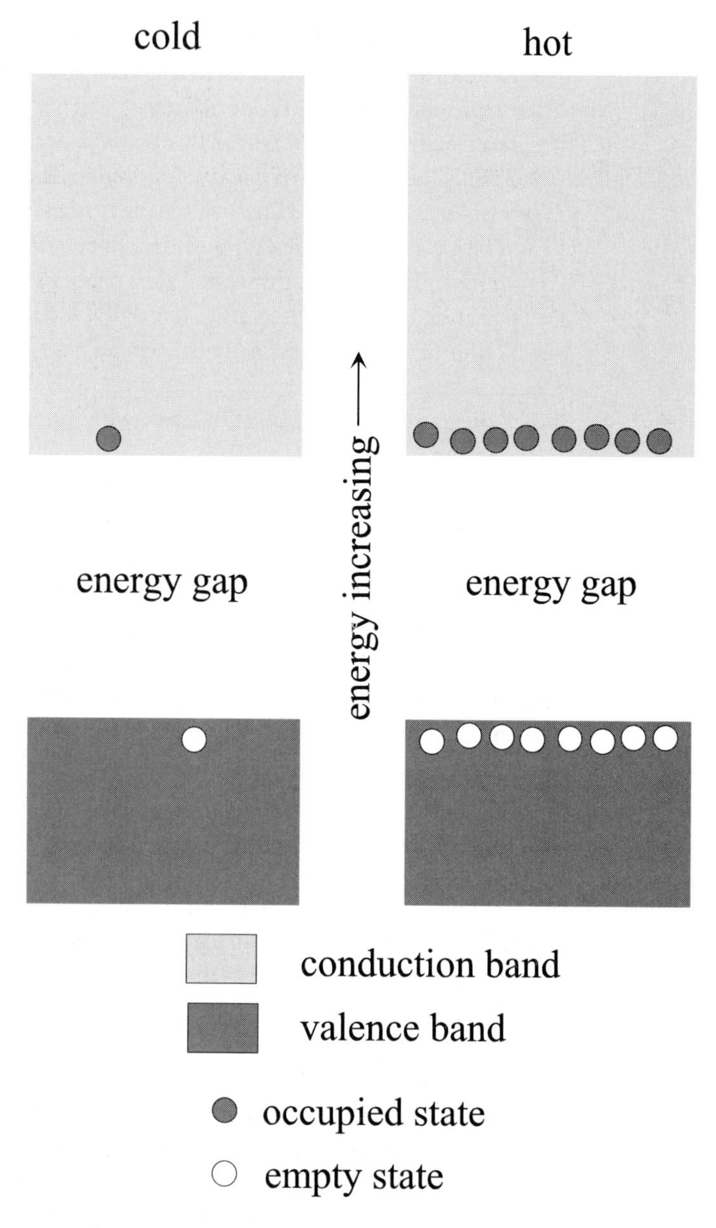

FIGURE X-4. The valence and conduction bands of a semiconductor at low and high temperatures. More electrons in the conduction band and empty states in the valence band (equivalent to a band of holes) are present when the semiconductor is hot than when it is cold.

An insulator, such as the material which covers the copper wire used in households for carrying electric currents, has essentially the same band structure as a semiconductor, but with one important qualitative difference. The energy gap between the filled band and empty band above it is so large, that at ordinary temperatures there are very few electrons with enough thermal energy to raise them from the filled band to the empty band. This means that negligible current is produced when a voltage is applied: we have an insulator.

We thus see that the main features of the electrical resistivity of metals, semiconductors and insulators can be understood in terms of the quantum energy levels of electrons and the quanta of thermal energy, namely the phonons. As noted earlier, the resistivities of different materials have the largest range of values of any of their physical properties. Nevertheless, the same basic quantum mechanical description applied to all these materials gives us an explanation of this tremendous variability. Quantum mechanics is not something which becomes important only when one is thinking of single electrons and atoms. It is essential also for the understanding of effects on an everyday scale, such as the flow of electric current through a wire.

5 Semiconductors at work

I suppose that everyone has heard of a *transistor*, though not all may be quite clear what it is. A small portable radio is often called a transistor, but this is not the meaning which was given to the term when it was invented. I shall use it here with its original meaning, to refer to an object made of a semiconductor so as to have certain desired electrical properties. All the electronic gadgets that surround us at home, at work and everywhere else contain such objects, ranging in number from a few to many millions. Examples are radios, television sets, video recorders, pocket calculators, automobiles, washing machines, computers, and on and on: one could make a very long list.

In order to give an example of what a transistor does, I consider a radio. It receives from the broadcasting station an electromagnetic wave which produces a very feeble varying electric current in the aerial. The radio converts this weak signal into a much larger current that flows through the loudspeaker and reproduces the broadcast. Transistors are the key element in this amplification of the signal. It is basically this function, of converting an incoming signal into a different outgoing signal, which is performed by the transistor.

Transistors come in many different forms depending on what they are intended for. I take the example of a type of transistor which the professionals call a MOSFET, an acronym for *m*etal-*o*xide-*s*emiconductor-*f*ield-*e*ffect-*t*ransistor. There are a million or more of these or similar transistors on a piece of silicon a centimetre or so across, at the heart of a computer. This is an example of the ubiquitous *chip*.

Figure X-5 is a sketch of a MOSFET. A section of p-type silicon is sandwiched between two n-type sections called source and drain respectively. A thin film of silicon oxide is sandwiched between the p-type silicon and a metal film which is called the gate. A voltage between the source and the drain should in general cause a current of electrons to flow from one to the other, since the source, being n-type, can supply electrons. The size of the current depends on how many electrons are available in the material between the source and the drain to carry it. This number can be changed by varying a voltage which is applied to the gate. Variations of this voltage (which could be the incoming signal in a radio, for example) are reproduced as variations of the current between the source and the drain, and there we have the basic functioning of a transistor. There is a similarity to a rubber hose through which water is flowing at a certain rate because of a pressure difference between its ends. This flow rate can be varied by pinching the hose at some point so that its cross-section changes there.

Other types of semiconductor devices are also in use, designed to fit particular applications. Fundamental to the working of all of them are two characteristics which are peculiar to semiconductors. The first is the special arrangement of the quantum energy levels into a valence band and a conduction band separated by an energy gap. The second is the ability to control the number and distribution in space and in energy of the electrons and holes by introducing junctions, by alloying, and by applying voltages.

6 Electric currents at work

The use of electricity is pervasive in the way of life today. In addition to the ever-increasing role of semiconductor devices, we use electric currents to produce motion, heat, light, and magnetism, to mention just a few of the applications. I describe now the basic effects of electric currents which make such applications possible.

The possibility of deriving motion from an electric current, as for example in an electric motor, depends on the fact that a wire carrying a current feels a force on it when it is placed in a magnetic field. I illustrate this in fig. X-6, which shows a current-carrying wire in the magnetic field between the north and south poles of a suitably shaped magnet. The wire feels a push in a direction which is perpendicular to the directions of the current and the magnetic field. Reversing the direction of either the current or the field reverses the direction of the push. By suitably arranging magnets and coils of wire and the way the current is fed to the coils, one ends up with a motor.

When an electric heater is switched on, current flows through it, and heat is given off. Suppose the current is I amperes, and the voltage is V volts. A current of I amperes means that I coulombs of charge, in the form of electrons, are flowing through the heater per second. The potential energy of the electrons is just the charge multiplied by

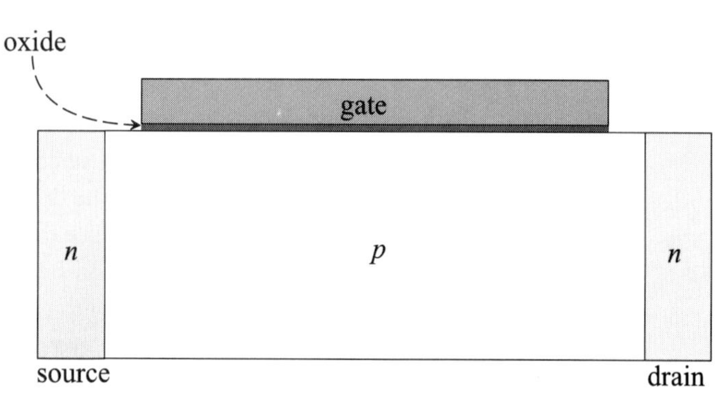

FIGURE X-5. A MOSFET. Changing the voltage on the gate changes the number of electrons available to carry a current in the part marked p, and this causes the current from the source to the drain to be changed accordingly.

the voltage. Since the voltage along the heater drops by V volts, the potential energy of the electrons leaving the heater per second is $I \times V$ joules less than that of the electrons entering the heater. This energy which is lost by the electrons in their passage through the heater appears as thermal energy, at a rate of $I \times V$ joules per second, or $I \times V$ watts. Now how exactly is the energy of the electrons converted to thermal energy of the heater coils? We saw that when a current is flowing in a resistance such as a heater coil, electrons are being constantly speeded up and then scattered back to states of lower energy, and the difference in energy appears as additional phonons. More phonons in

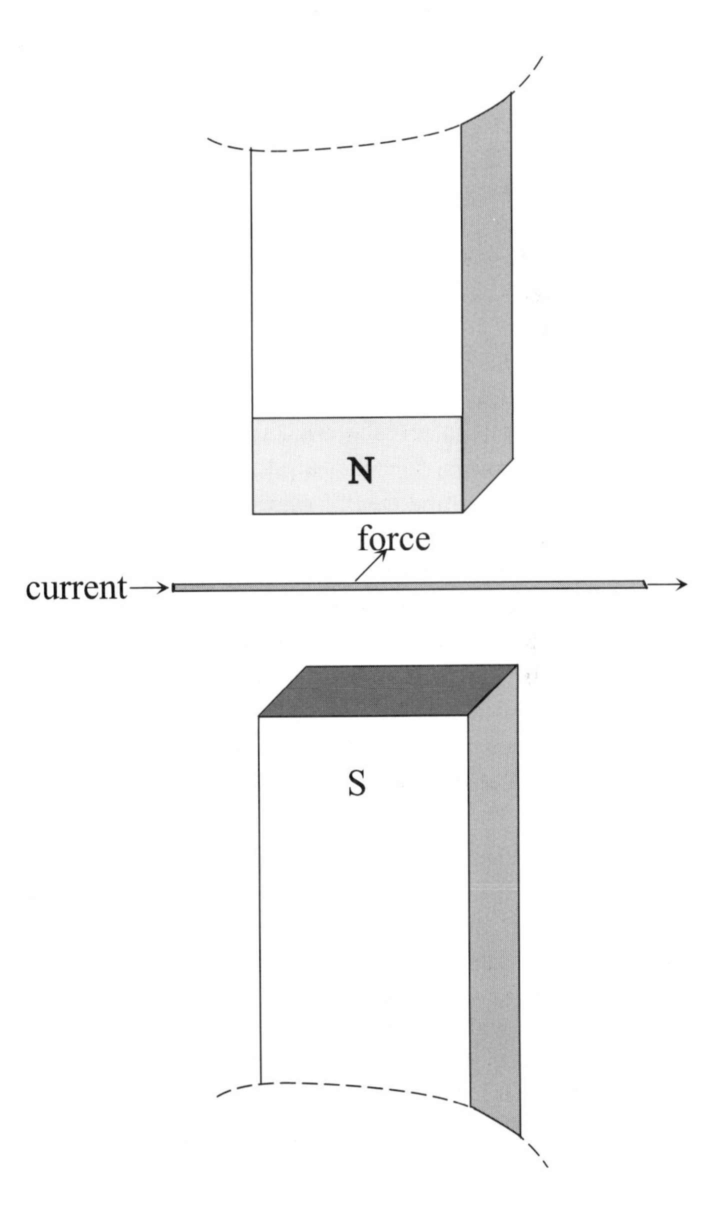

FIGURE X-6. Current-carrying wire in a magnetic field. The force is directed into the paper for the directions of current and magnetic field shown. Reversing either of these directions reverses the direction of the force also.

the wire means a higher temperature.

In an electric heater, the conditions are such that the energy produced as phonons is enough to heat the coils to a dull red heat. In an incandescent electric bulb, on the other hand, the current flowing through the filament causes it to become white hot, thus giving off light in addition to heat. Only a fraction of the electric energy supplied to a bulb appears as visible light; the rest just heats up the bulb and the outside. Fluorescent lights operate on a different principle. There, the kinetic energy of the electrons is absorbed by certain atoms which are therefore excited to higher energy states. The atoms then return to a lower energy state giving off the difference in energy as a photon of light. Such lights convert electrical energy to photon energy more efficiently than do electric bulbs.

We have seen that an electric current flowing in a wire produces a magnetic field around it, as can be seen by its effect on a compass needle. This property of a current is used to produce magnets consisting of a cylindrical coil of wire wrapped around an iron core and carrying a current. I shall discuss the magnetic properties of iron and some other materials in the next chapter.

7 Summary

The chapter is concerned with how an electric current is carried by metals and semiconductors, and why insulators cannot carry a current. I make an analogy in section 1 between the flow of electrons and the flow of water. I introduce the units used in measuring electric currents.

I describe in section 2 the property of a material called its electrical resistance, which is a measure of the ability to carry a current when a voltage is applied; the greater the resistance, the smaller the current for a given voltage. The resistance depends on the size and shape of the material through which the current is flowing. We define the resistivity of the material as the resistance of a one-cm cube of the material between a pair of opposite faces. One can then meaningfully compare the resistivities of different materials.

I note in section 3 that the resistivities of different materials vary over a tremendous range, the highest resistivities being more than 10^{18} times the lowest. The resistivity changes with temperature, and the variation is qualitatively different between metals and semiconductors. Alloying a metal increases its resistivity, and alloying a semiconductor changes its resistivity in a complicated manner.

In section 4, I describe the electrical current using the quantum mechanical picture of a solid. I find that the resistance in a pure metal or a semiconductor is produced by the scattering of electrons by phonons. The difference in the magnitude, and variation with temperature, of resistance in metals and in semiconductors is a consequence of their different energy band structures. I get a picture of how alloying changes the resistance.

I already described a solar battery, which is one practical application of the quantum mechanical picture of a semiconductor, in chapter IX. I give in section 5 a description of another example, namely a transistor. This is a basic unit in devices called integrated circuits (or chips) used in computers and many other things in everyday use.

I describe briefly in section 6 some practical applications of electric currents: to produce heat, motion, and magnetic fields. Electric currents occupy a central place in our daily lives.

We have so far looked at the thermal, optical and electrical properties of matter. We have seen how they follow from the atomic structure of matter as described by quantum mechanics. The two properties of electrons and nuclei which play a role in these properties are their mass and electric charge. We have not taken any notice of a third property that electrons, as well as most nuclei, have, namely that they are magnetic. They behave like permanent magnets which can be influenced by a magnetic field, just as their charge causes them to be affected by an electric field. We shall look in chapter XI at some consequences of this property.

XI MAGNETS

1 What is a magnet?

Magnets are a part of many everyday objects, although one is not always aware of their presence. You might be using one to hold your shopping list on the side of the refrigerator in the kitchen. The motor in the refrigerator contains a magnet, and its door is held shut by a magnet. Most household electronic equipment (radios, television sets and the like) contain magnets. Electric motors work because of the magnets in them. Magnets occupy a not insignificant place in the things used in our daily life.

I illustrate the basic properties of a magnet in fig. XI-1. When a magnet is mounted so that it is free to rotate, one end always turns towards the north: I have shaded this end in the diagram and labelled it N. I shall refer to this as the north pole and the other as the south pole of the magnet. The figure is divided into four parts labelled (*a*) to (*d*), and each illustrates a property of a magnet:

(*a*) When I bring two magnets together so that the north pole of one approaches the

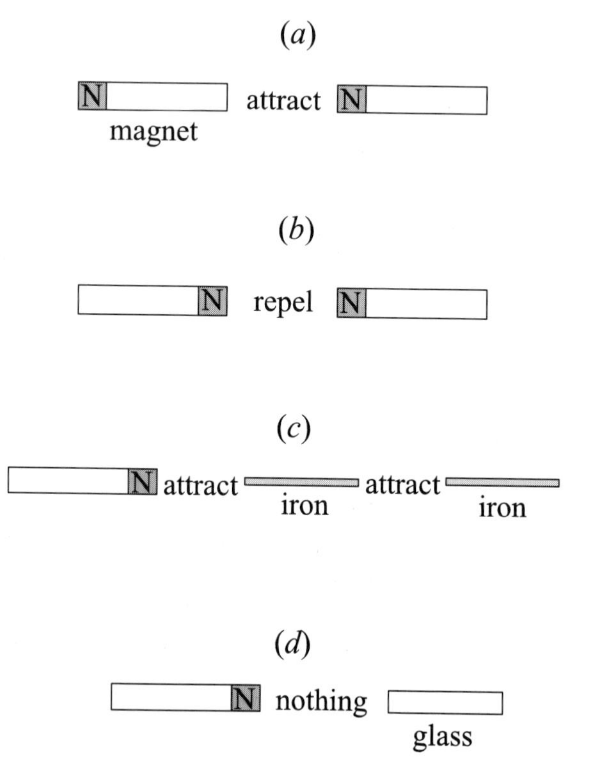

FIGURE XI-1. Basic features of a magnet. The north pole is marked N and shaded to identify it. (*a*) Unlike poles attract each other. (*b*) Like poles repel each other. (*c*) Either pole attracts a rod of iron, which then itself becomes a magnet capable of attracting a second bar of iron, which then itself becomes a magnet, and so on. The same happens if the rod is of nickel or cobalt or certain other materials. (*d*) The vast majority of materials, whether metals, semiconductors or insulators, feel practically no force from a magnet.

south pole of the other, the magnets attract each other: there is an attractive force between unlike poles.

(*b*) If two poles of the same kind approach each other, there is a force of repulsion between them.

(*c*) A rod of iron approaching either pole of a magnet is attracted towards it. Furthermore, the iron itself behaves like a magnet so long as it is close to the original magnet, and will attract another piece of iron. This induced magnetism of the iron vanishes when it is moved far enough away from the magnet. The magnet itself carries its magnetism all the time, which justifies calling it a permanent magnet to distinguish it from the iron rod. I get the same induced magnetism if I substitute cobalt, nickel, or certain alloys of these elements, for iron. Such materials are said to be *ferromagnetic*, and the phenomenon itself is *ferromagnetism*. If I heat the iron rod, I find that its ferromagnetism becomes weaker as the rod gets hotter, and finally vanishes when the rod is at 770 °C; the magnet exerts no force on the rod when it is above this temperature. The same effect is seen with other ferromagnetic materials, but at other temperatures.

(*d*) If the rod is of almost any other material (copper, glass, plastic, etc.), it feels a negligible force from the magnet. Such materials are not ferromagnetic.

We take the properties (*a*) and (*b*) as given by nature, rather like the forces between electric charges. In fact, it is tempting to push the analogy and to imagine that there are separate north and south poles, which could be called *magnetic monopoles*. The analogy here is with electrons and positrons which can be called *electric monopoles*. A magnet would then be a concentration of north poles at one end and south poles at the other end of the bar. However, many experimenters have tried without success to see if such magnetic monopoles actually exist. Our experience so far is that any magnet, whether of the kitchen variety or an electron or proton or neutron, always comes with both a north pole and a south pole. The two are inseparable: breaking a magnet in two just gives two smaller magnets, and not separated north and south poles. So it seems that an atomic description of magnetic materials must be based on the magnetism of the constituent particles as an immutable fact of nature, like for example the charge of the electron.

2 Elementary magnets

All matter is made up of electrons and nuclei, and just as with the other properties, we want to understand the magnetic properties in terms of these fundamental constituents of matter. I have said earlier that an electron, and most nuclei, behave as if they were spinning tops as well as permanent magnets. I go into this now in a little more detail.

I begin with the electron. It is a magnet, but a very weak one indeed, compared to an ordinary permanent magnet. I would need something like 10^{21} of these electron-magnets all pointing the same way and crowded into a volume equal to that of an ordinary bar magnet, in order to produce the same magnetic field as the magnet.

QUESTION Why must the electron-magnets all point the same way for this effect to happen?

ANSWER Otherwise, the magnetic fields produced at any point by the individual electrons would be pointing in all directions, and the net field at that point will be practically zero.

I measure the strength of a magnet using a unit called a *Bohr magneton*, named for the physicist Niels Bohr. The value of the Bohr magneton is fixed by the charge and mass of the electron and the Planck constant. It is therefore also one of the fundamental constants of nature. The electron-magnet is 1.001159652 Bohr magneton in strength, and also has a specific direction, namely from its south pole to its north pole. So I can represent the magnetism of the electron (or any other magnet for that matter) by a vector with magnitude and direction, and give it a name: *magnetic moment*.

A brief digression: you may have wondered at the extraordinary precision (one part in a thousand million) of the value quoted above for the electron-magnet's strength, its magnetic moment. It is the value found not only experimentally, but is also the value calculated by the combined theories of quantum mechanics, relativity and electromagnetism. It is a tribute to the extraordinary accuracy of some experimental work in physics, and a convincing proof that the physics used to calculate the value is correct.

The electron magnetic moments would tend to line up just like compass needles if I were to put them in a magnetic field. This tendency is opposed by their thermal energy which wants to make them point in all directions. The result at ordinary temperatures is that only a small fraction of the moments line up with the field. If the field is removed, even these moments become disordered. In a ferromagnet, the electron-magnets are in zero field. Nevertheless, they do line up and point in the same direction. This is the distinctive mark of a ferromagnetic substance and we shall see how it comes about later in this chapter.

The electron is also like a spinning top: it has a permanent angular momentum which is one of the constants of nature. The physicist Paul Dirac combined quantum mechanics and Einstein's relativity theory for the electron, and found that angular momentum and magnetic moment are like the two faces of a coin: one cannot have the one without the other. The quantization of angular momentum measured in a specified direction implies that the magnetic moment in that direction is also quantized. For the rest of this chapter, I shall talk only about magnetic moments, for we are concerned here only with magnetic properties.

The proton and the neutron are also magnets, but much weaker than the electron. It would take a couple of thousand of them to equal the magnetic moment of an electron. An atomic nucleus is made up of protons and neutrons, and so in general would also have a magnetic moment which is much smaller than a Bohr magneton. We shall see later that even these small moments are important in some cases.

QUESTION Is it possible for some nuclei to have zero magnetic moment?

ANSWER Yes. Suppose the numbers and the orientations of the proton- and neutron-magnets in some nucleus are such that the magnetic effects are cancelled pairwise. This will happen, for example, if all the protons and neutrons are separately paired up with the spins in each pair pointing in opposite directions, thus: ↑↓. Then the total magnetic moment of the nucleus will be zero.

I said that about 10^{21} electrons crowded together with all the moments pointing the same way would have the strength of an ordinary magnet, say a compass needle. This number is in fact comparable to the number of electrons in a piece of solid of that size. This spontaneous lining up of the electron moments takes place only in a few materials, among them iron, nickel and cobalt. Furthermore, we do not have to worry about the magnetism of the nuclei in this case, since it is thousands of times weaker than that of the electrons. How does this lining up of the electron moments take place? The answer is in the next section.

3 Ferromagnets

In an atom, most or all of the electron moments cancel out one another in pairs, with the two moments in each pair pointing in opposite directions. Then there may remain a few electrons whose moments have not got so cancelled out. Atoms with such unpaired electrons are potential candidates for forming ferromagnets. We know that there are forces of attraction and repulsion between magnets. So it is tempting to think that somehow these magnetic forces between the unpaired electrons on neighbouring atoms may lead to all the moments lining up, giving a ferromagnet. But when these forces are worked out in detail, they turn out to be too weak to give rise to the known ferromagnets like iron and nickel. We must look elsewhere for the origin of this property.

What we have ignored so far is the wave aspect of the electrons, and now we include this explicitly, as first done by the physicist Werner Heisenberg. Consider two neighbouring atoms in the solid, and suppose that each has one unpaired electron. Then quantum mechanics tells us that there is a wave function for the two electrons which gives the probability of finding each electron in any particular region in the two atoms. The quantum states (and the wave functions) for the two cases, of the two electron moments pointing in the same direction (parallel moments) or in opposite directions (antiparallel moments), are different. This leads to a difference in the potential energy due to the electric force between the two electrons in the two cases. What prevails in nature is the quantum state of lower energy. For iron and the other materials which are ferromagnetic, this lower energy obtains for parallel moments, while for all other materials it is the one with antiparallel moments. It is noteworthy that while ferromagnetism requires the magnetism of the electrons, it is not the magnetic force between them which leads to ferromagnetism, but rather the electric force and the wave-like nature of the electrons.

Now one might ask, if the moments in iron all like to line up in the same direction, then why is every piece of iron not already a magnet? Why does one have to bring up a normal magnet to the iron in order to make it behave like one? How is it that a permanent magnet comes already equipped with its magnetism, unlike the piece of iron? I shall answer these questions next.

The atoms in an iron crystal sit in a body-centred cubic structure, whose unit cell is shown in fig. II-13, part (4) (p. 28). The distance between neighbouring atoms is different in different directions, and one expects that the interaction between electrons on neighbouring atoms will also depend upon the direction. All this taken together leads to the result that the magnetic moments will point more easily along certain directions in the crystal, called the easy directions, than along others. In a lump of iron there are many very small crystals called grains which are all stuck together. These crystals point in all possible directions, and in each of them the moments point in one of the easy directions. So each grain is a magnet, but the different little grain-magnets point in different directions, and therefore the lump of iron itself has a total magnetism which is practically zero. I illustrate this in fig. XI-2, showing an enlarged view of the grains, each a magnet pointing a different way.

One might think that one would get around this problem by using a piece of iron which is entirely one crystal. There being no grains in it, the piece should have the electron moments all pointing in one easy direction, and therefore become a magnet. But this does not happen. The easy direction for the moments to point in iron is along an edge of the unit cube in the crystal: but there are six such directions in a cube, as you can see in fig. XI-3, part (*a*). The crystal gets divided into regions, and the moments in each of these regions will point in one of the easy directions, but they will be different in the different regions. Each of these regions is called a magnetic domain. I show in fig. XI-3, part (*b*), a possible arrangement of domains in a crystal shaped like a rectangular bar. The final result is the same as with the lump of iron: the single crystal bar will not show strong magnetism, because the effects of the different domains cancel one another out.

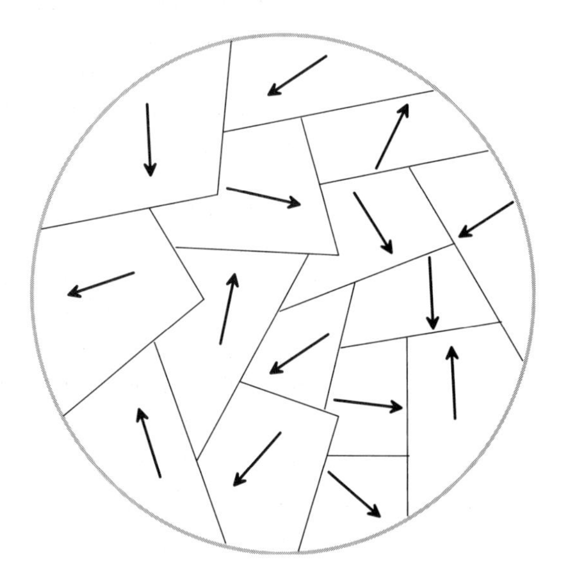

FIGURE XI-2. A microscopic view of a lump of iron showing its grains, each with its spins pointing in the direction of the arrow. These directions being different in the different grains, the lump as a whole shows no magnetism.

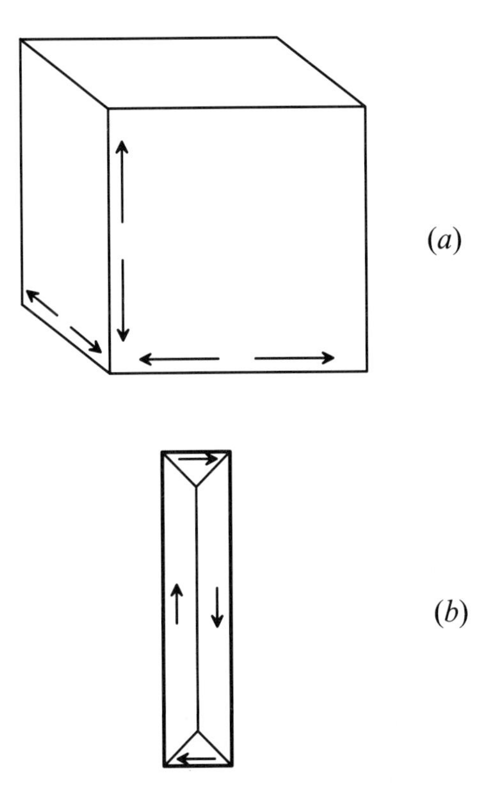

FIGURE XI-3. (*a*) The cube denotes the unit cell in the crystal of iron. The arrows along the cube edges show the six easy directions of magnetism in the crystal. (*b*) A possible arrangement of magnetic domains in a bar of single crystal iron. The arrows show the direction of magnetism in each domain. The bar as a whole is then only weakly magnetic at best.

QUESTION Why does a small enough ferromagnetic particle, e.g. a grain, have only one domain?

ANSWER The total energy of a ferromagnetic body includes contributions from the energy of the magnetic field outside the body and the energy of the boundaries between adjacent domains. These two energies have different dependences on the size of the body, such that a large body prefers it (i.e. has lower energy) when it forms many domains, whereas a small enough particle prefers to be a single domain.

QUESTION Why should the domain structure in fig. XI-3, part (*b*), produce no magnetic field outside the bar?

ANSWER Consider the two long domains which are side by side with the arrows pointing up and down respectively. They are like two opposing magnets side by side, and at any point in the space around the two will produce fields which are nearly equal in size but in opposite directions, thus cancelling each other out. Similar cancellation occurs with other pairs of domains, and so the bar as a whole will produce a negligible field outside.

But we still have to explain permanent magnets, and why one of them makes a piece of iron behave like a magnet when the two are brought together. A small enough ferromagnetic particle prefers to have just one domain rather than several, because then its energy is lower. A permanent magnet contains such single-domain ferromagnetic particles. These particles are suspended in a substance which itself need not be ferromagnetic. During the manufacturing process, there is a stage when the particles are free to rotate. This is when a magnetic field is turned on, causing the particles to rotate so that the magnetism in each particle is pointing in the direction of the field. As the process continues, the substance surrounding the particles hardens and they become frozen in their new orientations. The result is a permanent magnet, a microscopic view of which is shown in fig. XI-4. Another process starts with the particles alone, without the surrounding substance. They are aligned in a magnetic field as before, and then heated to a high enough temperature so that they stick to one another and are no longer free to move. This way of consolidating particles is called sintering.

And now to the bar of iron which becomes a magnet when it is brought near a permanent magnet. What happens here is that, if the magnetic field of the permanent magnet is strong enough, the magnetism in each of the domains of the iron will rotate to an easy direction which is closest to the direction of that field. The bar then behaves like a magnet. When the permanent magnet is removed, the domains return to their random orientations, and the magnetism of the bar as a whole is lost. I am reminded of a classroom in which, before the teacher comes, the pupils are clustered in small groups with each group talking about a different topic. The teacher enters, and all eyes and attention turn to her, and there is only one topic, hers. When she leaves, the initial condition is, at least usually, restored.

One can also use the magnetic field produced by an electric current flowing through a spool of copper wire wrapped around a rod of iron to make it magnetic, and then one has an *electromagnet.* This is the kind of magnet used in many applications where magnetic fields are needed.

Finally, I consider how the magnetism of iron and other ferromagnetic materials changes with temperature. The ferromagnetism becomes weaker as the temperature of

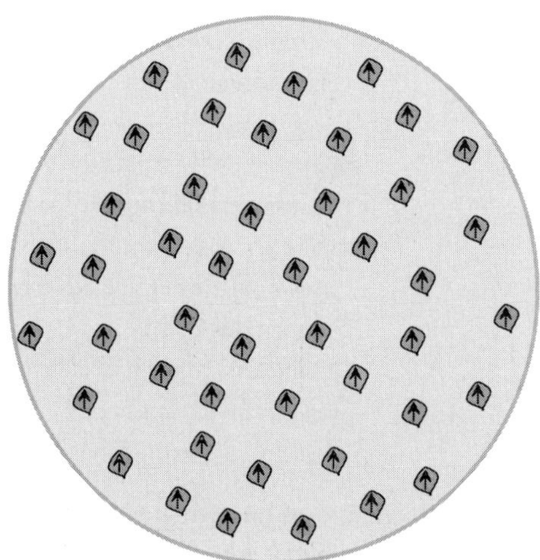

FIGURE XI-4. A microscopic view of the magnetic particles in a permanent magnet. Each particle is a single domain, and the magnetism, shown by an arrow, points in the same direction for each domain.

Table XI-1. *Curie temperatures (in kelvin)*

material	Curie temperature (K)
iron	1043
nickel	631
cobalt	1403
gadolinium	289
dysprosium	105

the material is raised, and finally vanishes at a well-defined temperature which is called the Curie temperature after the physicist Pierre Curie. Table XI-1 shows the Curie temperatures of several ferromagnetic substances. Note that the temperatures are in kelvin, so that 0 °C is 273 K. The Curie temperature varies from well below room temperature to very high temperatures, depending on the material.

Figure XI-5 shows how the strength of the magnetism varies as the temperature changes from near the absolute zero to the Curie temperature. It has the highest value at the lowest temperature, gradually drops as the temperature is raised, and becomes zero at the Curie temperature.

QUESTION Can this effect be because the electron gradually loses its magnetic moment as the temperature goes up?

ANSWER No. The electron's magnetic moment, like its mass or charge, is an intrinsic property which does not change with temperature.

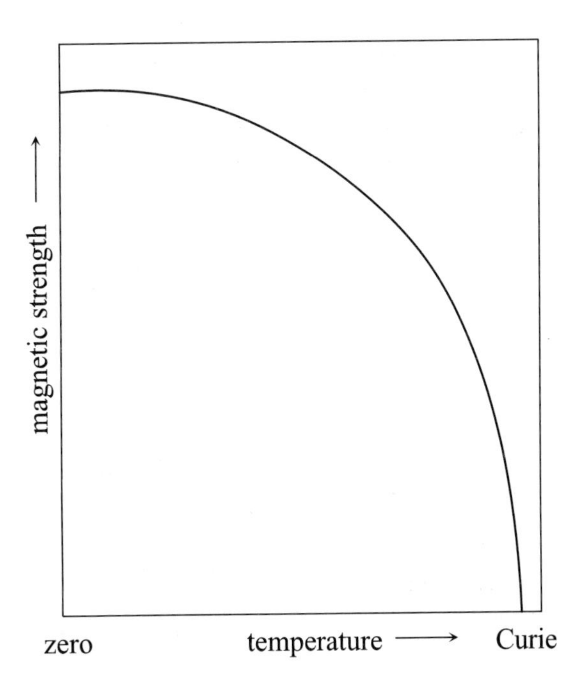

FIGURE XI-5. Variation of magnetic strength with temperature for a ferromagnetic substance.

The reason for the decrease in the magnetic strength is that with rising temperature more and more of the electron moments begin to point in the opposite direction, until at the Curie temperature there are as many moments pointing one way as there are the opposite way, and the net magnetic strength then becomes zero. This loss of ferromagnetism on going through the Curie temperature is an example of what is called a *phase transformation.* One gets the ferromagnetic phase below the Curie temperature, and it transforms into a non-magnetic phase on being heated above that temperature. There are other phase transformations in nature, and trying to understand how they come about has been, and continues to be, one of the challenging problems in physics. I shall introduce you to this topic in the next section.

4 Phase transformations

I describe a few examples of phase transformations, from which we shall be able to see some essential common features. There are two examples from our daily experience: the melting of ice to water at 0 °C, and the boiling of water to steam at 100 °C. In each case, the change occurs at a sharply defined temperature. The atomic arrangements are totally different in the phases below and above the transformation temperature. Consider ice: even if the temperature is just a little below 0 °C, it is all ice, and just a little above, all water. The change does not take place one molecule at a time, so to speak, but all at once. Ice is a crystal: there is order in the arrangement of the molecules. This order suddenly disappears when ice melts to water. I can talk about order in a ferromagnet also, measured by the net fraction of moments pointing in one direction. This order does not disappear abruptly at the Curie temperature, but decreases smoothly and finally vanishes at that temperature. I list some common features of these transformations:

1. There is a sharply defined transformation temperature, which I denote by T_0.
2. There is more order in the phase below T_0 than in the one above T_0. The actual type of order can be different in different cases: crystalline order in melting, the ordered orientation of electron magnetic moments in ferromagnetism.
3. The change in this order in going through T_0 can be smooth, as in a ferromagnet, or jumpy, as in ice.
4. The change affects all the atoms (or electrons or molecules, as the case may be). Ice does not melt one molecule at a time.

The physical properties of the two phases are dramatically different, and yet all it takes to go from one to the other is ever so small a change in temperature. So you can see that there must be subtle features of the statistical physics of the constituent particles (the molecules in ice, the electrons and atoms in a ferromagnet) which must be carefully considered in understanding such phase transformations.

I shall devote chapter XII to a famous example of a phase transformation, namely superconductivity. It is one of the best understood transformations, and provides us a splendid opportunity to put together the different strands of the quantum mechanical description of solids which we have learnt so far.

5 Nuclear magnetic resonance

It was all right to neglect the magnetism of the atomic nuclei in considering ferromagnetism, because the magnetism of the electrons, which is the source of ferromagnetism, is thousands of times stronger. However, there is another difference between the two which gives nuclear magnetism some features absent in electron magnetism. Magnetically considered, all electrons are identical, but all nuclei are not. Different nuclei have different magnetic moments and spins, as table XI-2 shows.

Table XI-2. *The magnetic moment of various nuclei*

nucleus (isotope number)	magnetic moment (in nuclear magnetons)	spin (S)
hydrogen(1)	+2.792743	$\frac{1}{2}$
carbon(13)	+0.702381	$\frac{1}{2}$
oxygen(17)	−1.89370	$\frac{5}{2}$
phosphorus(31)	+1.13162	$\frac{1}{2}$
copper(63)	+2.22664	$\frac{3}{2}$

The isotope number shown in parenthesis is the total number of protons and neutrons in the nucleus. The nuclear magneton is a unit which is smaller than the Bohr magneton by a factor of about 5.4×10^{-4}. The magnetic moments differ in magnitude for different nuclei. They can be positive or negative, which indicates whether the nuclear magnetic moment points in the same direction as the angular momentum vector or in the opposite direction.

The angular momentum of a nucleus is specified by the spin quantum number S, whose values are also shown in the table. When I put the nucleus in a magnetic field, its energy changes owing to the action of the field on the magnetic moment. Because of quantum mechanics, we are not surprised to find that this change is quantized. The energy of the nucleus can only take one of a certain number of discrete values, and nothing in between. Even the number of these allowed quantum energy levels is not arbitrary: it is equal to $2S+1$, where S is the spin quantum number. This is because there are only $2S+1$ specific directions in which the magnetic moment can point with respect to the field. We have here one more example of how the quantum nature of matter leads to results which are quite different from our everyday experience. It is as if a compass needle, instead of pointing north, could only point, say, either northeast or southeast, and in no other direction.

There are thus two levels in hydrogen ($2S+1 = 2$ if $S = \frac{1}{2}$), six in oxygen, and so on. You see here one consequence of the intimate connection between magnetic moment and angular momentum. For a given nucleus, the energy difference is the same between each pair of adjacent levels, and is directly proportional to the strength of the magnetic field. This energy difference also depends on the magnetic moment of the nucleus, and is therefore different for different materials, even when the magnetic field is the same. I illustrate these features in the energy level diagrams in figs. XI-6 and XI-7.

Suppose I put some hydrogen in a magnetic field, so that each nucleus has two energy levels differing by an amount of energy E. I now shine photons of frequency f on the hydrogen. Each of these photons has energy hf, where h is the Planck constant. A nucleus which is in its lower energy level can absorb a photon and move to its higher energy level only if the photon energy is exactly equal to the difference in energy between the two levels:

$$E = hf.$$

If this condition is not satisfied, the photon cannot be absorbed and will just move right through the material. The absorption, as the frequency of the photons is varied, is shown in part (*a*) of fig. XI-8. There will be no absorption except at the frequency f: we speak of an absorption peak at that frequency.

Now suppose I take some material which contains several kinds of atoms, and place it in a magnetic field. Because of the different nuclear magnetic moments, each set of

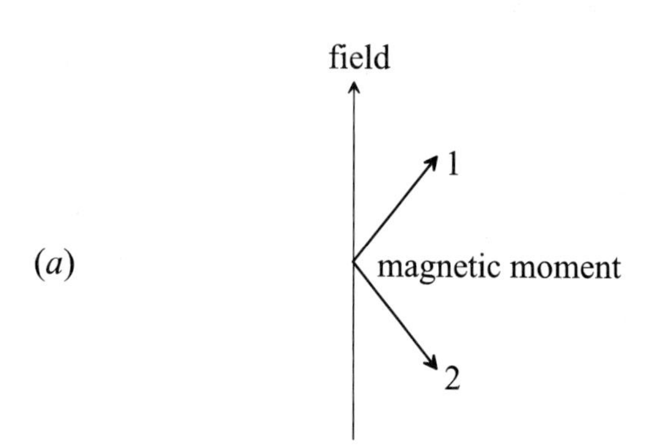

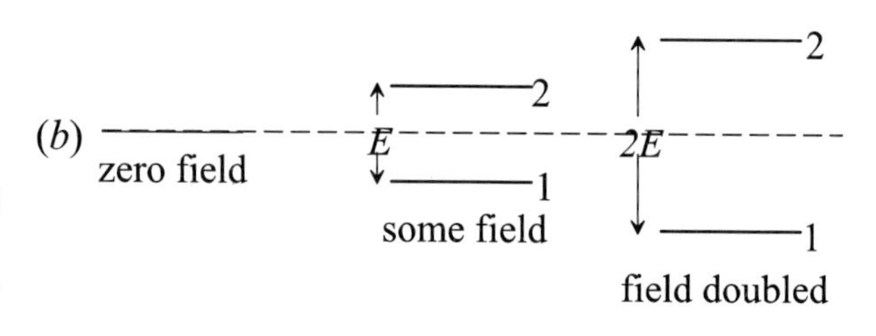

FIGURE XI-6. Effect of a magnetic field on a nucleus of spin $\frac{1}{2}$. Its magnetic moment can only point in one of two directions labelled 1 and 2, but not in any other direction, as shown in part (*a*) of the figure. Part (*b*) shows the quantum energy levels for these two orientations. The difference in energy between these levels increases from E to $2E$ when the field is doubled.

FIGURE XI-7. The energy levels in the same magnetic field of the nuclei of hydrogen (which is just a proton), carbon, and oxygen. The spin quantum numbers for these nuclei are respectively $\frac{1}{2}$, $\frac{1}{2}$, and $\frac{5}{2}$. Note that the energy differences between levels are different for different nuclei.

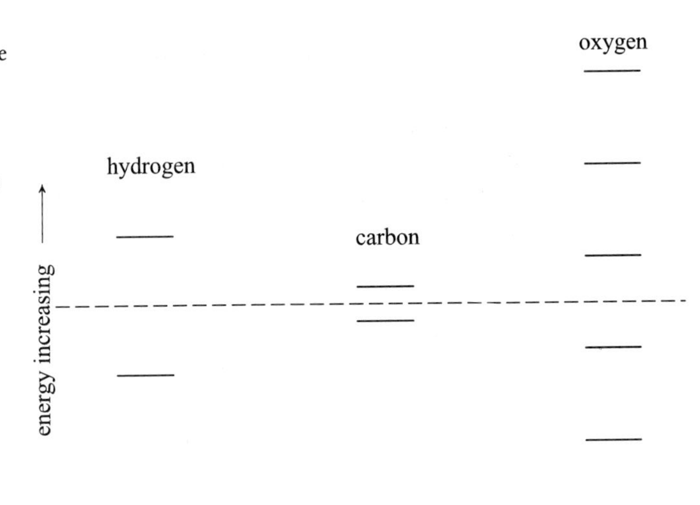

FIGURE XI-8. Nuclear magnetic resonance. Part (*a*) illustrates the sharp rise in absorption of energy from an electromagnetic wave at a certain frequency, by nuclei placed in a given magnetic field. Part (*b*) shows the case of two different kinds of nuclei, hydrogen and carbon, say, where the frequency is constant and the magnetic field is varied. Each peak in the absorption arises from one of the two types of nucleus.

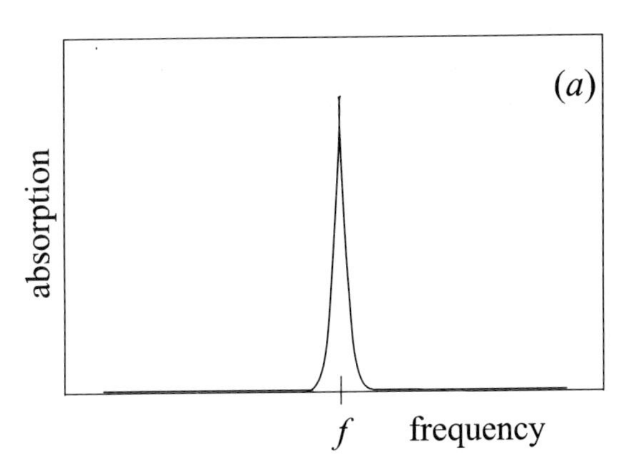

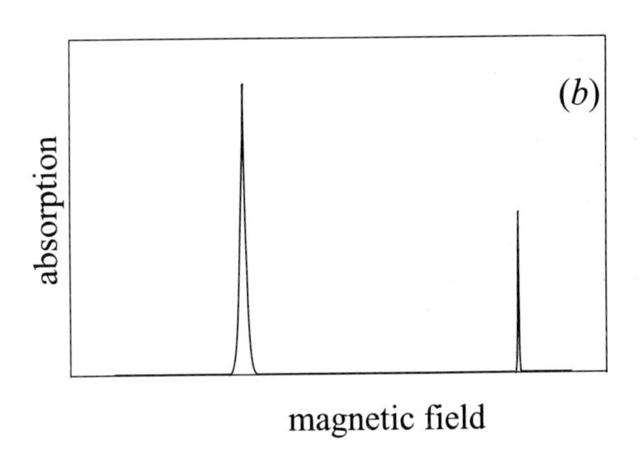

identical nuclei will set up its own distinctive group of energy levels, different from that of any other nucleus. If I now shine photons of a single frequency on the material and increase the magnetic field gradually, I get a strong absorption from each type of nucleus at that value of the field for which its energy spacing is equal to the photon energy. I show in fig. XI-8, part (*b*), what the absorption might look like for a mixture of two different types of atoms at fixed photon frequency and changing magnetic field. In practice it is much easier to keep the photon frequency constant and to vary the magnetic field, and so this is what is usually done.

This absorption of photons by nuclei placed in a magnetic field is called *nuclear magnetic resonance.* It is at the heart of magnetic resonance imaging (MRI), a technique used in medical diagnostics. Since each distinct type of nucleus carries its own signature in the form of a resonance signal, it is possible to make a picture of the inside of the human body point by point, as it were, without the unwelcome necessity of invasive procedures.

The electron of course also has a magnetic moment, and so one would expect to see similar resonance effects due to electrons. They are indeed observed, and the phenomenon is called electron spin resonance. Both types of resonance find many applications in the study of materials, living as well as inanimate.

6 Weak magnetism

I have considered so far only ferromagnetic materials; but these form only a very small minority of all the materials we have. All the others have electrons too: the electrons in shells around the nuclei, and in addition electrons which are free to move about in the case of metals. The electrons in shells have in general an angular momentum, and therefore a magnetic moment too, which is in addition to the intrinsic magnetic moment of the isolated electron. A free electron in a metal, when it sees a magnetic field, feels a force which is always perpendicular to both its direction of motion and the direction of the magnetic field. The net result is that the electron finds itself going round in a circle or a helix. This in turn is just like a loop of wire carrying an electric current, and therefore it produces its own magnetic field. An electron going round in a circle or helix thus has a magnetic moment which is in addition to the intrinsic magnetic moment arising from its spin.

When one of these materials is brought into a magnetic field, I get a spectrum of energy levels because of the effect of the field on the various types of magnetic moments I have described. These levels are occupied by the electrons in accordance with the Pauli principle. Each level corresponds to a different magnetic moment. When they are all added up, most of them cancel one another out, and what is finally left is a very weak net magnetic moment, much smaller than what we find in a ferromagnet. All materials which are not ferromagnetic show this kind of weak magnetism.

7 Summary

I consider in this chapter the consequences of the magnetic properties of electrons and nuclei. The one which is closest to everyday experience is ferromagnetism, as for example in iron. I describe in section 1 the basic properties of ferromagnets.

Section 2 deals with the magnetic properties of electrons, and introduces the magnetic moment as a measure of the strength of a magnet. The magnetic moment is intimately connected to the spin (or equivalently the angular momentum) of the electron.

I take up ferromagnetism in section 3. I find that the electrons' magnetic moments line up in these materials not, as one might have expected, because of the magnetic force between them. Rather, it is via the electric force between the charges combined with the nature of their wave function: a purely quantum-mechanical effect. I find the reason why the strength of a magnet varies with temperature. I explain why an ordinary piece of iron is not a permanent magnet.

In section 4 I describe what is meant by the term *phase transformation.* Examples are the melting of ice to form water, and the change of iron from a weak magnetic material to a ferromagnet as it is cooled through its Curie point.

I go in section 5 to the magnetism associated with nuclei. I explain the phenomenon of nuclear magnetic resonance, which is at the heart of the magnetic imaging systems used in medical diagnostic work.

Section 6 deals with materials which do not show ferromagnetism, but which nevertheless become weakly magnetic when placed in a magnetic field. This magnetism, which vanishes when the field is removed, is also caused by the quantum energy levels of the electrons which are produced by the effect of the magnetic field on their motion and their magnetic moments.

I introduced the idea of a phase transformation in section 4. I take up in the next chapter a spectacular example of such a transformation, namely the total disappearance of electrical resistance when certain materials are cooled to low temperatures. This is the phenomenon called superconductivity.

XII SUPERCONDUCTORS

1 What is a superconductor?

I have noticed that physicists are rarely given to excesses of exuberance in talking about their subject, even among themselves. So when they use a prefix like *super* for a material, it must be something quite extraordinary. I shall first tell you what the phenomenon is, and then explain why it is extraordinary. Suppose I take a wire of the metal lead which has a resistance of one ohm at room temperature, about 290 K, and measure its electrical resistance at different temperatures as I cool it down. I show the result in fig. XII-1. The resistance drops smoothly as the temperature drops. Near 7 K, the resistance drops abruptly by a factor of more than 10^{20} below its value at room temperature, and remains so at all lower temperatures. Put differently, the wire now conducts electric currents at least 10^{20} times better than it did before. In fact, we can only set an upper limit to this number because of the limits on the sensitivity of the measuring instruments. For all we know, the wire in this condition may be a *perfect* conductor, which means that a current can pass through it with zero voltage applied to it. Well, *super* is a bit less than *perfect*, and so we are content to say that lead has become superconducting, and to call the phenomenon itself *superconductivity*.

If I now warm up the lead wire from the superconducting state, I find that its resistance jumps back to its original value at the same temperature at which the resistance had disappeared when cooling the wire. I call this temperature the *superconducting transition temperature*, and use the symbol T_0 (read T-nought) to denote it. When the wire is above this temperature and therefore not superconducting, I say that it is in the *normal state*. The change from one state to the other is very abrupt: a change of temperature of less than a thousandth of a degree is enough to cause the transition.

We begin to see why superconductivity is so extraordinary. There is the enormous change in the electrical resistivity, and then there is the abruptness (meaning the narrowness of the temperature range) with which it occurs. We saw in chapter IX how electrical resistance arises. It involves interactions between the conduction electrons, phonons, and alloying atoms: a complicated interplay of huge numbers of particles (more than 10^{20} electrons alone, not to mention the others). This cacophony is abruptly and totally stilled just by cooling the wire a thousandth of a degree through the transition temperature. Something happens which affects the interplay of all these particles, and which dramatically changes their behaviour. Here is an analogy: I hear a din emerging from a classroom, and it suddenly ceases when a bell rings. Subsequent research shows me that the reason for the silence was the

FIGURE XII-1. Superconductivity of lead metal. Note that the scales are logarithmic, and not linear. This is a convenient way to show the variation of quantities over a large range of values. Thus here the temperature varies by a factor of a few hundred, and the resistance by a factor of 10^{20}. The resistance drops precipitously to zero at a temperature of about 7 K.

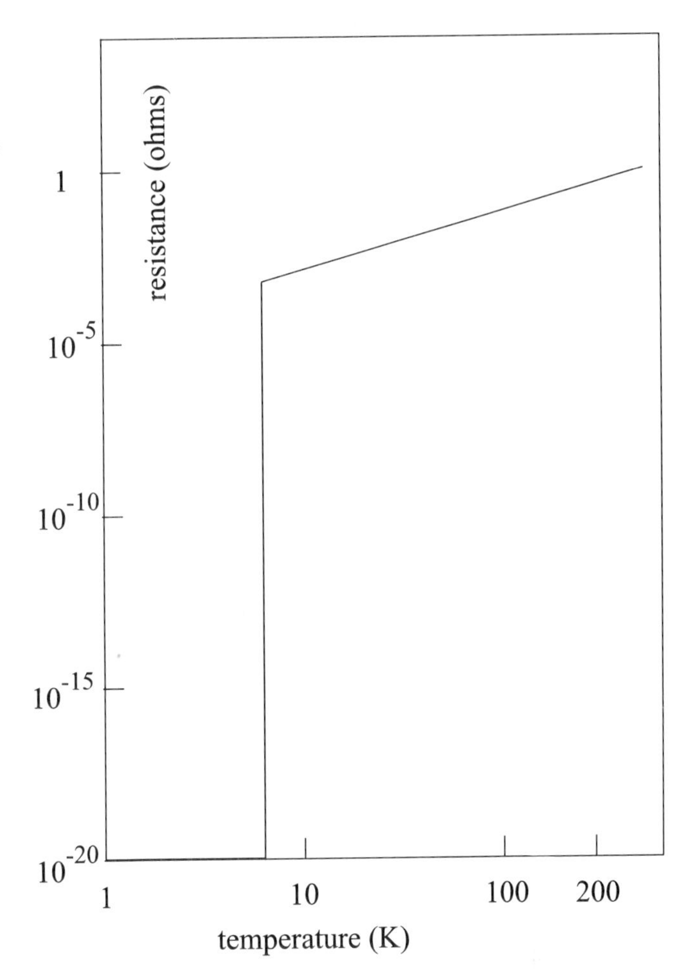

entrance of the teacher. So the task is to find out what plays the role of the teacher in superconductivity.

We are accustomed to seeing small changes in properties with small changes in temperature: the thermal expansion of solids, for example. This implies that the fundamental structure of the material, meaning its basic atomic and electronic structure, also undergoes correspondingly small changes. The situation is quite different with superconductivity. The abrupt disappearance of electrical resistance implies that some new feature must have appeared in the superconducting state which just did not exist in the normal state. This is an example of a *phase transformation.* Ferromagnetism and the melting of solids are other examples of such transformations. They are all characterized by a sudden change in the group (as distinct from individual) behaviour of all the relevant particles in the material: the conduction electrons in a superconductor, the electron magnetic moments in a ferromagnet, the atoms in a solid undergoing melting. Because of the large numbers of particles involved and the complicated interactions among them, it is a formidable challenge to explain them in detail. The effort to do so has been most successful with superconductivity, and I shall describe it later in this chapter.

Superconductivity is, perhaps surprisingly, anything but a rare occurrence in materials. In addition to lead, there are twenty-seven other naturally occurring elements which become superconducting, each with its own characteristic transition temperature. There are hundreds of alloys and compounds which are also superconducting. Superconductivity is a far more widespread phenomenon than ferromagnetism, which appears in only a few materials.

Table XII-1. *Some superconducting transition temperatures*

superconductor	transition temperature (K)
tungsten	0.012
tin	3.7
lead	7.2
an alloy of niobium and titanium	11
a compound of niobium and tin	18
a compound of yttrium, barium, copper and oxygen	90

I list in table XII-1 the transition temperatures of a few superconductors. I have selected them to bring out two points: the variety of materials which can become superconducting, and the great range of transition temperatures. The two extreme temperatures in the table differ by a factor of almost eight thousand. I should mention here as an aside that in physics, a comparison of two temperatures is often more meaningful in terms of their ratio than their difference. This is because in statistical physics, which is at the heart of the idea of temperature, the ratios of the energies of quantum levels to the thermal energy (which is basically what temperature is) are more significant than the temperature by itself. The percentage change in this ratio is the same in going from, say, one degree to ten degrees, as from a hundred degrees to a thousand degrees.

QUESTION Even though superconductivity is much more prevalent than ferromagnetism, why is it that I see so many uses of magnets around me and none of superconductors?

ANSWER Well, at least part of the answer lies in a comparison of tables XI-1 and XII-1. We have materials which are ferromagnets at ordinary temperatures (around 300 K, say) but none which are superconducting at these temperatures. Any use of superconductors would require refrigeration with liquid air (90 K) or liquid helium (4 K) using special equipment, and this rules out the everyday use of superconductors on a scale anywhere near that of magnets. The search continues for materials which are superconducting at room temperature, so far without success.

2 Some other properties of superconductors

The disappearance of electrical resistance is the most spectacular property which distinguishes a superconductor from a normal metal. A matter of terminology: I use the term *normal* in this chapter as synonymous with *non-superconducting*. Thus lead is normal above 7.2 K, and superconducting below that temperature.

Now imagine that I have the wire of lead sitting at a temperature of around 4 K, which is the same as minus 269 °C. I get to such low temperatures by using liquified helium gas. The wire has an electric current flowing through it but with no voltage, because it is superconducting. Can I do anything to the wire to disturb its superconductivity? Shaking it or shining light on it does nothing. Knowing that electric current is carried by electrons, and that a magnetic field influences their motion, I let a magnet approach the wire. At first nothing happens; the current continues to flow undiminished. But when the strength of the field at the wire has reached some apparently critical level, the current in the wire suddenly drops and a voltage appears: the wire has lost its superconductivity and now shows its normal electrical resistance. I try a few more tests, and find that at each temperature, there is a critical strength of magnetic field, below which the lead wire is superconducting and above which it is normal. I show in fig. XII-2 how this critical strength varies with temperature, beginning with very small values near T_0 and levelling off at a maximum of about 800 gauss at the lowest temperatures. The gauss is the unit in which the strength of a magnetic field is measured. The strength of the earth's field is about half a gauss. Different superconductors show the same qualitative variation of critical field as the temperature is varied, but have different values for this maximum critical field, for which I use the symbol H_M. Table XII-2 lists H_M for several superconductors. One sees that superconductors fall into two groups in terms of H_M. There are those whose critical fields are a few hundred gauss or less, and then those with fields of hundreds of thousands of

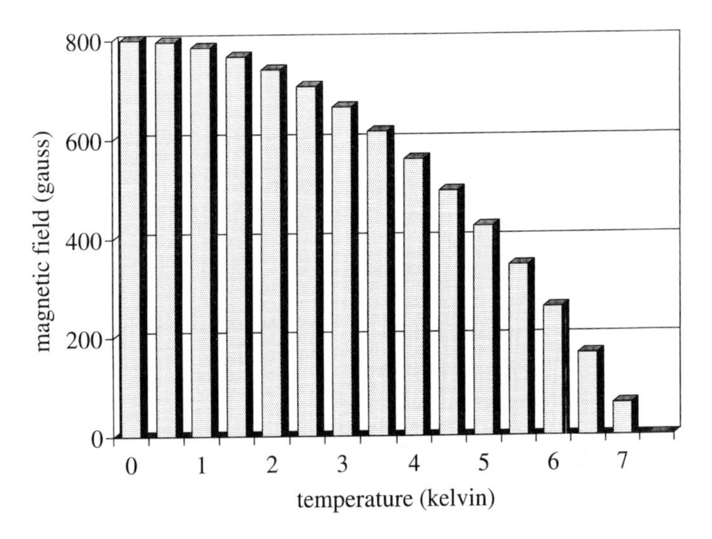

FIGURE XII-2. The smallest magnetic field (in gauss) which will destroy superconductivity in lead at different temperatures. Fields less than this do not affect the superconductor.

Table XII-2. *Maximum field H_M for superconductivity in some materials*

superconductor	maximum critical field, H_M, in gauss
tungsten	1
tin	300
lead	800
an alloy of niobium and titanium	120,000
a compound of niobium and tin	300,000
a compound of yttrium, barium, copper and oxygen	estimated to be more than 1,000,000

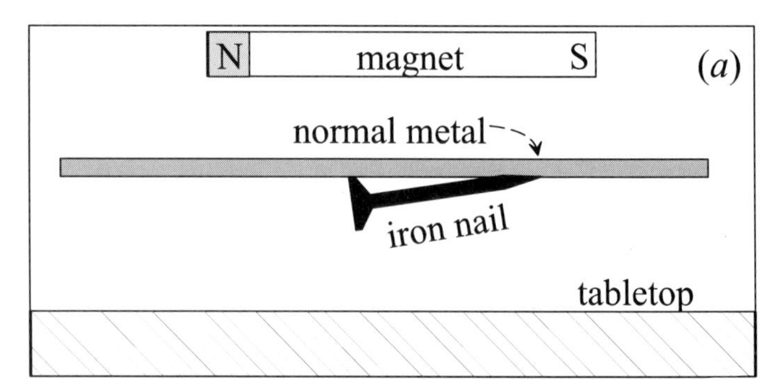

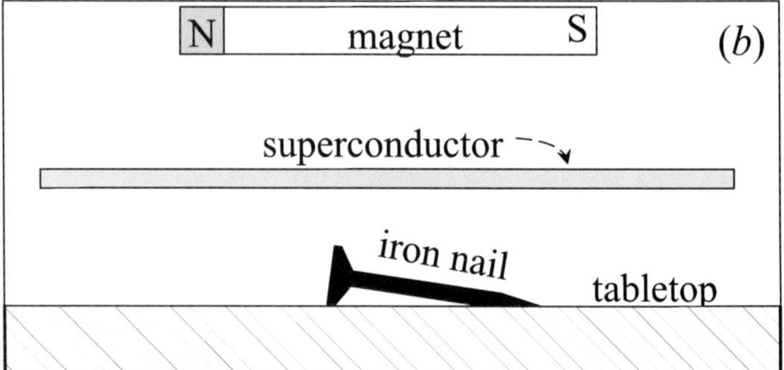

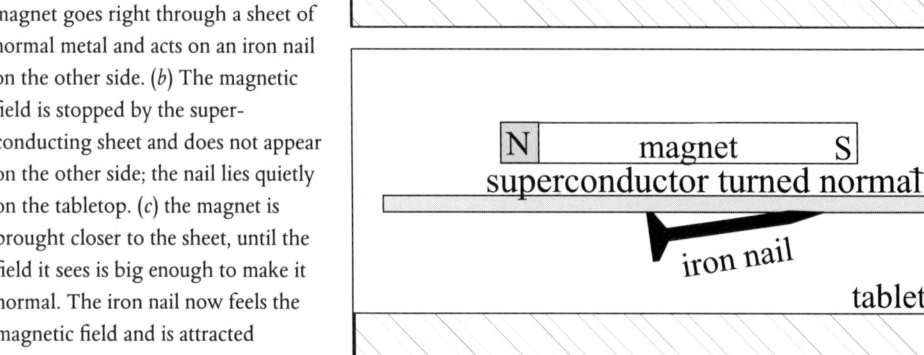

FIGURE XII-3. (*a*) The field of a magnet goes right through a sheet of normal metal and acts on an iron nail on the other side. (*b*) The magnetic field is stopped by the superconducting sheet and does not appear on the other side; the nail lies quietly on the tabletop. (*c*) the magnet is brought closer to the sheet, until the field it sees is big enough to make it normal. The iron nail now feels the magnetic field and is attracted towards the sheet.

gauss. I shall refer to them as the low-field (LF) and the high-field (HF) superconductors respectively. The reason for the different behaviours lies in the details of how the superconductor reacts to a magnetic field, as we shall see now.

A magnetic field goes right through a material which is not superconducting: a magnet will attract an iron nail even if I interpose a sheet of normal metal in between, as shown in fig. XII-3, part (*a*). If I replace the normal metal by a superconducting sheet, as in part (*b*) of fig. XII-3, the magnetic field is stopped by it: the iron nail is not attracted by the magnet. If now I bring the magnet gradually closer to the superconducting sheet, I find that at a certain point the magnetic field apparently punches through the superconductor, because now the nail is attracted towards the magnet, as in part (*c*) of fig. XII-3.

QUESTION Can you identify one essential detail which is missing in fig. XII-3?

ANSWER Depending on what it is made of, the superconductor must be sitting in a bath of liquid air or liquid helium, because we have no materials which are superconducting at room temperature.

Further experimenting shows that the magnetic field punches through the superconductor if it is the LF type when its strength is equal to H_M, which is also the field at which the material loses its superconductivity. I illustrate this effect in fig. XII-4. In part (*a*) of the figure, the field is less than H_M, and cannot break into the superconductor (shown as an egg-shaped object in the picture). This is known as the Meissner effect, for Walther Meissner who in collaboration with R. Ochsenfeld discovered it. In part (*b*), the field is larger than H_M, and becomes the same inside and outside the object, which is now normal. It is as if the superconductor tries to keep the field out because that is how it can remain superconducting, and the excluded field exerts a sort of pressure to get into the material. This pressure increases as the field increases, until finally the superconductor can no longer stand it. The field then punches its way in and kills the superconductivity.

On the other hand, if the object is an HF superconductor, its behaviour is more cunning, so to speak, than that of the LF type. I illustrate it in fig. XII-5. In small fields, part (*a*), the HF type excludes the field just like the LF type. But at some medium field which is much less than H_M for that material, it makes a compromise and lets a part of the field in, as shown in part (*b*) of the figure. The material then becomes an intimate mixture of superconducting and normal regions, and is said to be in the *mixed state.* As the field increases, it admits a larger and larger fraction of the field, and the normal fraction of the volume grows at the expense of the superconducting fraction. Finally at H_M the field inside and outside the material become equal, as shown in part (*c*), and the material becomes normal throughout its volume.

The two kinds of magnetic behaviour described above go under the names of type I and type II behaviour respectively. All HF superconductors show type II behaviour, and they are the ones which are mostly used in practical applications.

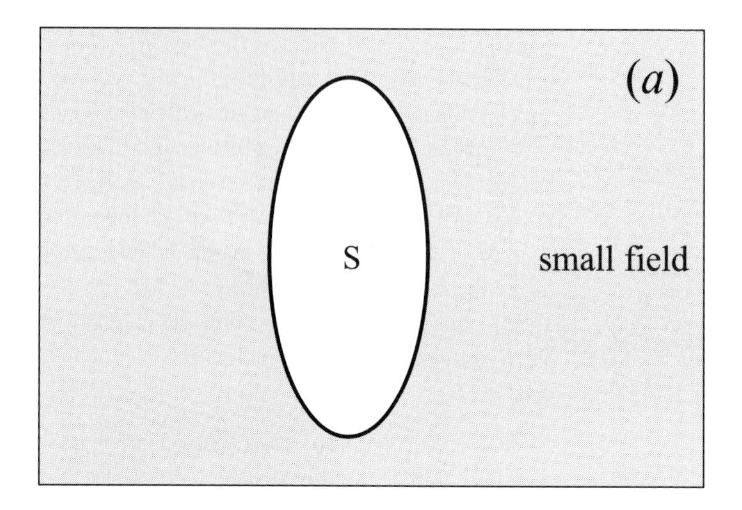

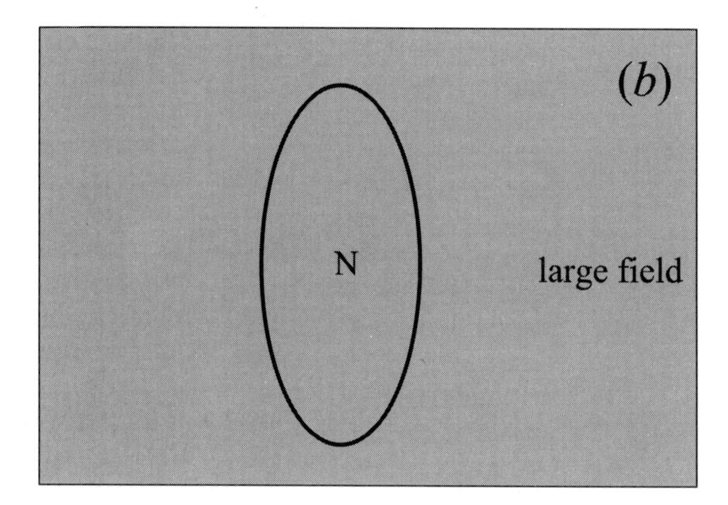

FIGURE XII-4. A type I superconductor, shown as an egg-shaped object, (*a*) keeps a small field out, and (*b*) lets a large field in. The changeover is at the field H_M, at which superconductivity ends. The state of the sample is indicated: S = superconducting, N = normal.

In a normal metal, I need to apply a voltage in order to send an electric current through it. In a superconductor, there is no resistance and a current can flow even without a voltage. If I start a current flowing in a superconducting ring, for example, it will continue for ever without diminishing, because there is no resistance. I call it a *persistent current.* This current going round the loop produces a magnetic field: it has a magnetic moment. This, like everything else in nature, is described by quantum mechanics. The magnetic moment is quantized. It can take only certain discrete values, and nothing in between. Measurements of this quantization give an unexpected result: the particles in the persistent current are not electrons acting singly and separately, but rather pairs of electrons acting as a unit. This is quite different from the normal state of the metal, in which the properties are those expected from electrons behaving singly. It is as if I made experiments with oxygen gas, and found that the

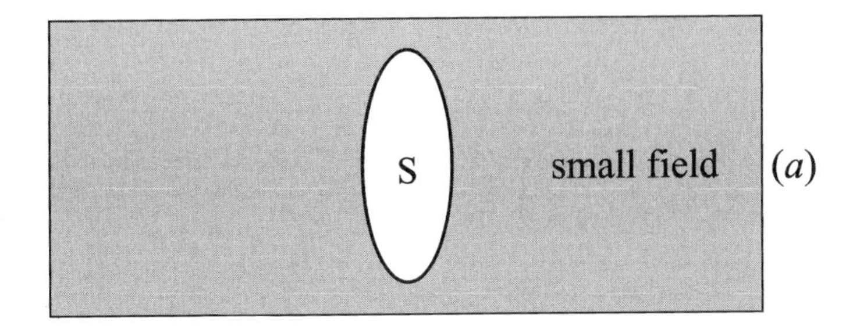

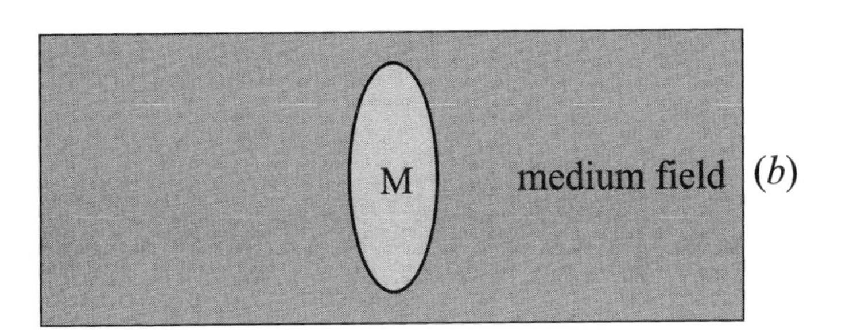

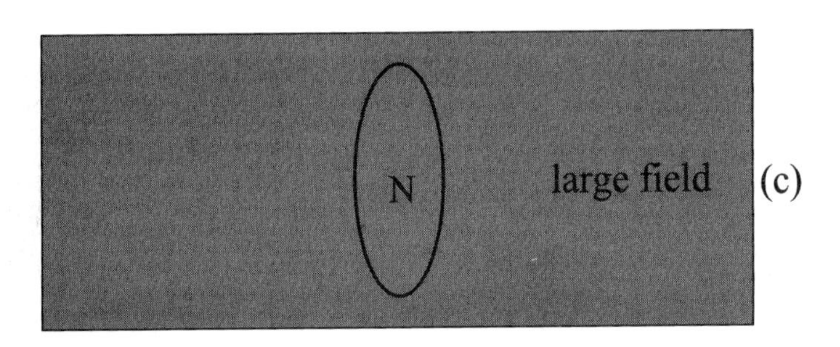

FIGURE XII-5. (*a*) A type II superconductor, shown as an egg-shaped object, keeps a small enough field out. (*b*) Intermediate fields leak partially into the superconductor. (*c*) When the field H_M for the particular material is reached, the field enters completely into the object and its superconductivity ends. The state of the sample is indicated: S = superconducting, M = mixed, N = normal.

results made sense only if the gas consisted not of single oxygen atoms, but rather of pairs of atoms bound together.

Some of the properties in the superconducting state of a material are the same as in the normal state, and others are quite different. The density and crystal structure do not change. The thermal conductivity and the specific heat, on the other hand, are quite different in the two states. The thermal conductivity does exactly the opposite of what we might expect from our understanding of normal metals. We saw there that a metal which is a good electrical conductor is also a good conductor of heat: copper for example. We might therefore expect that a superconductor would be also a super thermal conductor. Not so at all; the thermal conductivity of a metal in the superconducting state is smaller than what it is in the normal state. In the metal lead, for example, the thermal conductivity near 0 K in the superconducting state is about a million times smaller than in its normal state. The metal, although electrically a superconductor, becomes thermally an insulator. Since heat in metals is conducted primarily by electrons, it would appear that though the electrons in a superconductor carry an electric current with no resistance at all, they are strangely reluctant to carry a heat current. The specific heat in the superconducting state also is less than in the normal state.

3 Some clues

The explanation of how superconductivity comes about is one of the most impressive achievements of physics. It represents a beautiful bringing together of several aspects of the physics of solids that I have described earlier in this book: the quantum mechanics of electrons in a solid, phonons, the nature of electrical conduction, magnetic effects. Let me list some basic facts about superconductivity, and indicate how they may suggest where to look, or for that matter where not to look, for an explanation of the phenomenon. This is not the historical sequence; I wish rather to give you the picture as it is today.

1. In contrast to ferromagnetism, superconductivity is a rather common occurrence: there are hundreds of known superconductors. They come in many different crystal structures and chemical compositions. The only thing they have in common, apart from their superconductivity, is that they are metallic conductors above their transition temperatures. So we must seek an explanation in some features which are quite general in such conductors, and do not depend specially on crystal structure and chemical composition.

2. The transition temperatures are mostly well below 100 K. We recall that temperature is a measure of the thermal energy of the system. The average kinetic energy of the conduction electrons in a metal, expressed as a temperature, is around 100,000 K. Thus the energies involved in superconductivity are a minute fraction of the total electronic energies. The needle-in-a-haystack image springs to one's mind.

3. The average energies of phonons are not too far from the energies corresponding to the transition temperatures. One might therefore wonder whether phonons should be somehow brought into the picture. This thought is reinforced by the fact that the transition temperatures of different isotopes of the same substance differ slightly, and we know that the isotopes differ slightly in their phonons, but not in their electronic structure.

4. The fact that electron pairs are the units which carry the supercurrent suggests that we should look for something which will be the glue, as it were, binding a pair of electrons together.

5. We recall that heat is conducted in metals predominantly by the conduction electrons and holes which have been raised to slightly higher energy states near the Fermi energy because of their thermal energy, as illustrated in fig. XII-6. The poor thermal conduction in a superconductor suggests that fewer electrons get to these states

than in a normal metal. This could happen is if there is a gap between the Fermi energy and the energy of the lowest available state for electrons of thermal energy, as shown in fig. XII-7. If such a gap opened up when the metal becomes superconducting, then there would be fewer electrons to carry heat in this state than there are in the normal state, and the superconductor would be a poor thermal conductor and also have a lower specific heat, as is observed.

QUESTION With reference to the third point above, why should the isotopes differ slightly in their phonons?

ANSWER The isotopes have different nuclear, and hence atomic, masses. The phonons are the quanta of the vibrational energies of the atoms in the solid, and these energies are different for different atomic masses.

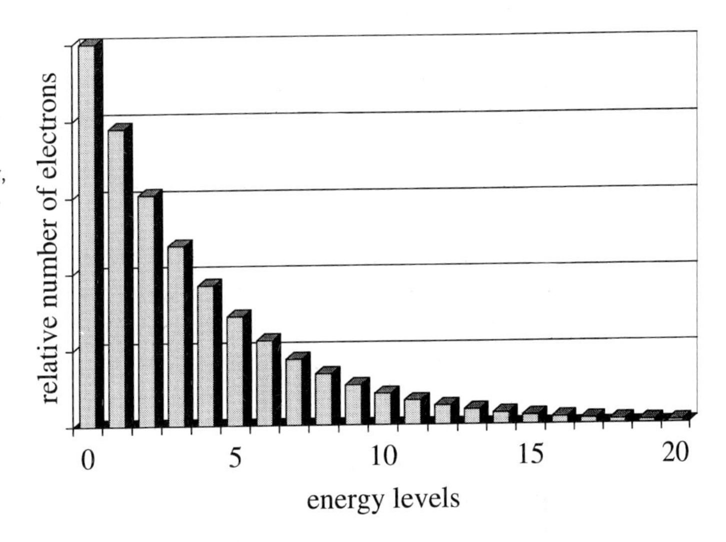

FIGURE XII-6. Relative numbers of electrons in different states of energy above the Fermi energy in a normal metal at some temperature. The level labelled 0 is nearest the Fermi energy, the others are at progressively higher energies.

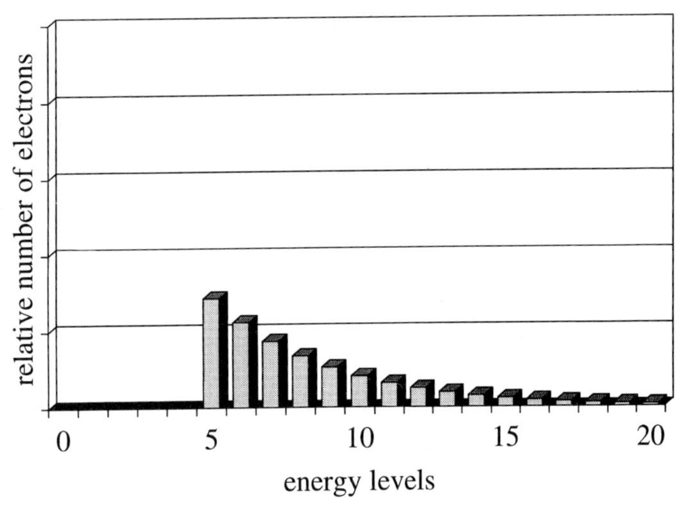

FIGURE XII-7. Relative numbers of electrons in different states of energy above the Fermi energy in a superconductor at some temperature. The lowest allowed energy for an electron is now in the level labelled 5. The lower energy levels are prohibited: we have an energy gap. By comparing with fig. XII-6, one can see that there are fewer thermally excited electrons, and therefore a lower thermal conductivity and lower specific heat in the superconducting state than in the normal state.

4 Cooper pairs and BCS

The physicist Leon Cooper showed how a certain kind of interaction between electrons can lead to the formation of electron pairs composed of opposite momenta and spin, which have therefore come to be called *Cooper pairs.* A Cooper pair is a peculiar object. It is not two electrons somehow bound to each other so that they stay close together and move about as a unit: an electronic molecule consisting of just two electrons, so to speak, just as an oxygen molecule consists of two closely bound oxygen atoms moving about. In fact, it is not possible for two electrons by themselves to be bound together: the electric force of repulsion between their charges would make them fly apart. One needs a mediator between them to hold them together in spite of this repulsion. An example of such a mediator is the nucleus of the helium atom, which holds two electrons bound close to each other. This is because each electron is attracted to the positive charge of the nucleus, and this attraction more than overcomes the repulsion between the electrons. I can talk about an electron pair in the helium atom bound to each other, and it takes a finite energy to break up the pair: this would be the energy needed to remove one electron and leave behind a helium ion.

The mediator that holds the two electrons of a Cooper pair together is a phonon, which is the quantum of energy of the thermal vibrations of the atoms in the solid. I illustrate how this comes about in fig. XII-8. Two electrons A and B approach each other travelling in opposite directions with the Fermi velocity. Electron A emits a phonon and moves off in a different direction. Electron B absorbs the phonon and changes its direction of motion by the same amount as A, so that the two electrons are still moving in opposite directions. The pair remains intact, bound together, so to speak, by the phonon. This is a Cooper pair. This process happens to all pairs of electrons moving with opposite velocities all over the Fermi surface, and the result is the superconducting state. Remembering that momentum is mass times velocity, we can say that a Cooper pair is two electrons of opposite momenta (and therefore of zero net momentum) bound together by the emission and absorption of a phonon. The two electrons also have opposite spins: this is established through experiments and predicted by the theory.

QUESTION How does an electron emit or absorb a phonon?

ANSWER The negative charge of the electron pulls towards it the positively charged ions surrounding it, causing a local increase of pressure. This is also what a sound wave would do, and so one can speak about the emission of a phonon by the electron. A half-cycle of the sound wave later, what was a compression has become an expansion of the lattice. If now a second electron comes along, it pulls (through the electric force) the surrounding ions back to their normal positions, and this corresponds to the absorption of the phonon.

We now have a picture of how Cooper pairs are formed through the mediation of phonons. But there is more to it than just the formation of pairs: it turns out that the

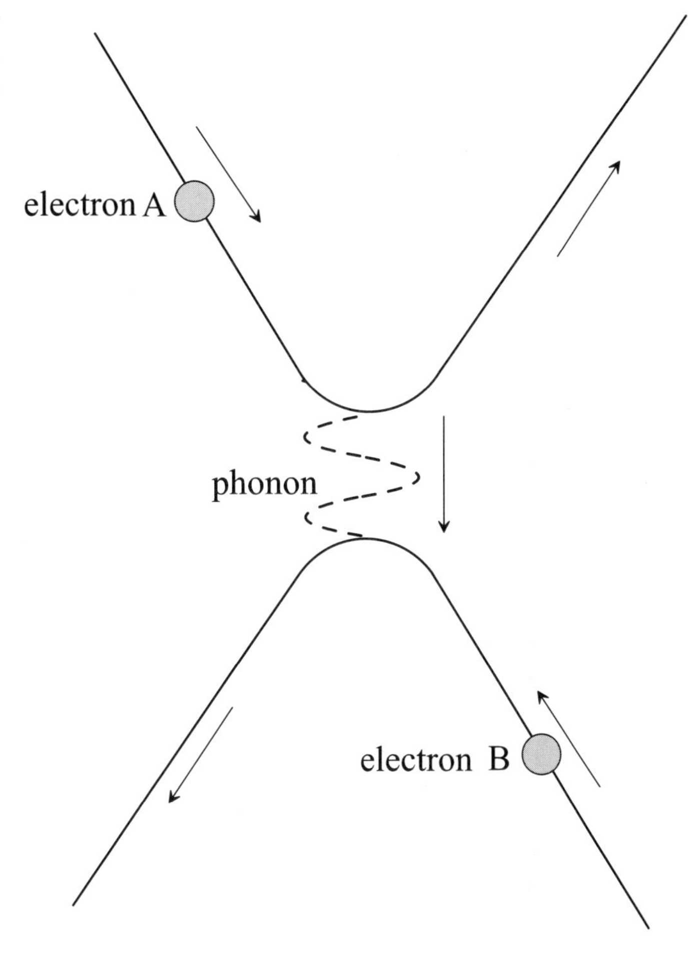

FIGURE XII-8. Formation of a Cooper pair. Electrons A and B approach each other travelling in opposite directions. Electron A emits a phonon which is absorbed by B, and they both change their directions of motion which are still opposite to each other.

energy of the electrons in such a pair is less than the energy that the electrons would have if they were moving independently. When electron A creates and emits a phonon, what is happening is that it is pulling towards itself the positive ions in its neighbourhood because of the electric force between its negative charge and their positive charges: you remember that a phonon represents a compressional wave in the solid. The potential energy of the electron is therefore lower than it would be if the ions were in their undisturbed positions. Electron B, which arrives there half a cycle of the phonon later, sees these ions being further from it than would otherwise be the case, and therefore its energy is raised. However, this increase is smaller than the decrease in energy of electron A. The net energy of the Cooper pair is thus lower than that of the two electrons when they are moving independently. The difference between the two is the binding energy of the Cooper pair, and must be supplied to it in order to break up the pair.

I have now described how one Cooper pair is formed. All the electrons in the vicinity of the Fermi surface are paired up in this manner at the absolute zero of temperature in a superconductor. You will notice that the formation of a pair involves the two electrons going from an initial pair of states to a final pair through the emission and absorption of a phonon, and this is only possible if the final pair of states is not already occupied by electrons.

> QUESTION Why is it necessary for the final states to be empty?
>
> ANSWER Because electrons are fermions, and no more than one electron can be in any given state. If there is already an electron in that state, then a second electron cannot enter it.

The answer above means that Cooper pair formation depends on what all the other electrons are doing: there is strong correlation among the Cooper pairs, they are not independent entities. I give a rough analogy: the electrons in a normal metal are like the molecules in a gas, they are all moving about independently. The electrons in a superconductor are like the molecules in a crystal. These molecules are all strongly correlated with one another, so that only the crystal can move as a whole, and not the single molecules independently of one another. Just as we talk about the molecules in a gas condensing to form a crystal, we have a sort of condensation of the electrons in the superconductor into a new state which is the superconducting state. The analogy is only rough: the molecules in a crystal condense into an ordered state with respect to their positions, and this is not the case with the electrons in a superconductor. They condense into an ordered state with respect to their momenta. But there is still something left in the analogy. The atoms in the gas have momentum because they are moving about, whereas the atoms in the crystal are sitting still (apart from their thermal vibrations) with zero momentum. Similarly, the electrons in a normal metal are all moving about, whereas each Cooper pair in the superconductor has zero momentum and is effectively sitting still. The crystal can be moved only as a whole, and each molecule in it has exactly the same motion as the crystal itself. One of the molecules can have a different motion only if it is pulled out of the crystal, and this requires a finite amount of energy. Similarly, all the Cooper pairs in the condensed state move as a whole, and to pull single electrons out of it requires a finite amount of energy, namely the binding energy of a pair.

Figure XII-9 suggests what the essential difference is between the normal and superconducting states of a metal. You must imagine as you look at the figure that the individual electrons in all cases are speeding about, but the Cooper pairs are stationary (with zero current) or have a common velocity (with current). In the normal metal, I need in principle to keep track of all the electrons to understand its behaviour. In the superconductor, however, I need to know the behaviour of just one Cooper pair: all the others will be doing exactly the same thing.

We can now see why an electric current in a superconductor flows without resistance. The current is carried by the condensed Cooper pairs which are all moving at the same velocity. Electrical resistance is caused by the scattering of electrons moving independently of one another. Such electrons can only be pulled out of the Cooper pair condensate when at least a finite minimum amount of energy is supplied to it. This

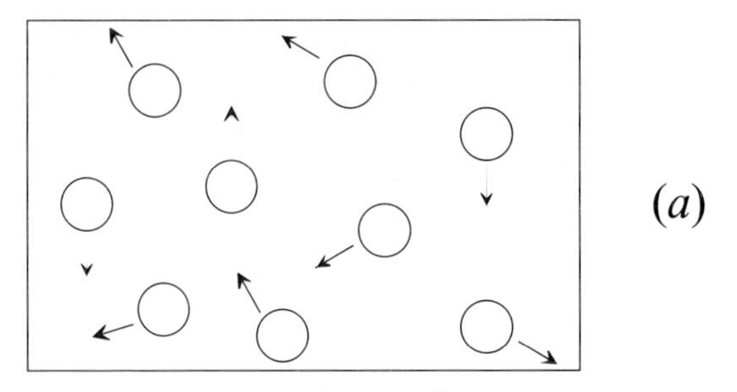

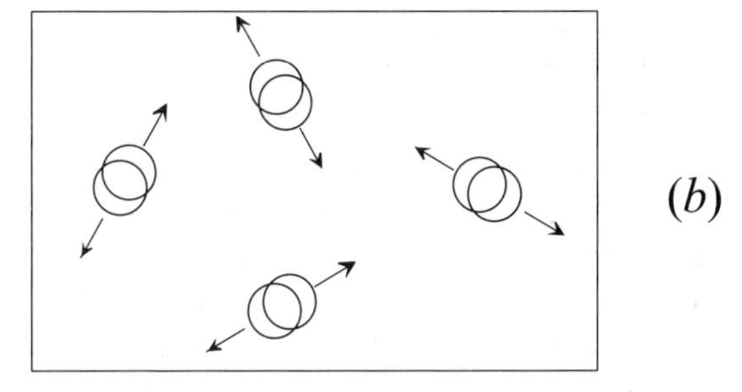

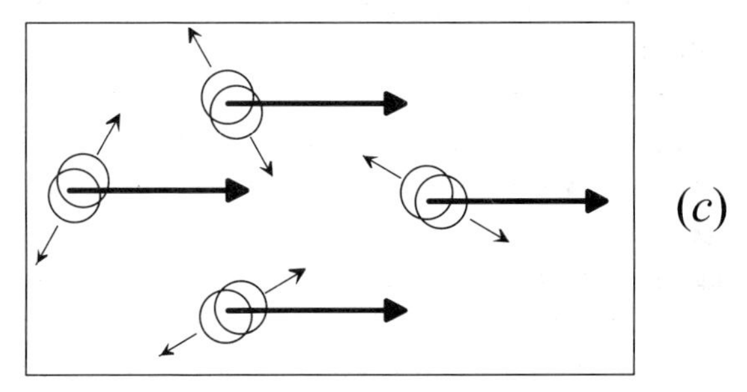

FIGURE XII-9. In a normal metal, part (*a*), electrons are moving about independently in all possible directions. In a superconductor, parts (*b*) and (*c*), Cooper pairs are formed by electrons with opposite velocities and spins. The total momentum of a pair is zero, so that its effective velocity is also zero, when there is no electric current, as in part (*b*). They all move with the same velocity in the presence of a current, as in part (*c*). The short arrows indicate directions of motion of individual electrons, and the long arrows show the direction of motion of each Cooper pair when there is a current.

energy would, for example, be available in the kinetic energy of motion of the pairs when the electric current is large enough. The energy could then go into breaking up the pairs, and normal resistance would be restored. This provides a mechanism whereby the superconductor can carry only a specific maximum current, called its *critical current*, before it loses its superconductivity and becomes a normal metal with resistance.

The picture of a superconductor which I have described above was proposed by the physicists John Bardeen, Leon Cooper and Robert Schrieffer, and is known in the world of physics as the BCS theory. They, and others following them, showed that by working out its consequences in detail, one could explain all the observed properties of superconductors. The BCS theory is one of the great achievements of the quantum mechanical description of nature. It was originally developed to show how the large number of electrons in a metal can influence one another through the mediation of phonons and thereby form the superconducting state. But its significance is far broader, because it showed a way of looking in general at fermions which are interacting with one another. It has been used, for example, to explain properties of atomic nuclei, where the fermions are protons and neutrons, and of certain astronomical objects called neutron stars, composed of very densely packed neutrons.

5 Josephson oscillations

The physicist Brian Josephson used the BCS theory to predict an effect one should see in superconductors. The effect was indeed found experimentally, and is named after him. Two superconducting pieces A and B are brought together with a thin insulating layer C interposed between them as shown schematically in fig. XII-10, thus forming a junction. A battery is connected to A and B, so that there is a voltage of V volts between

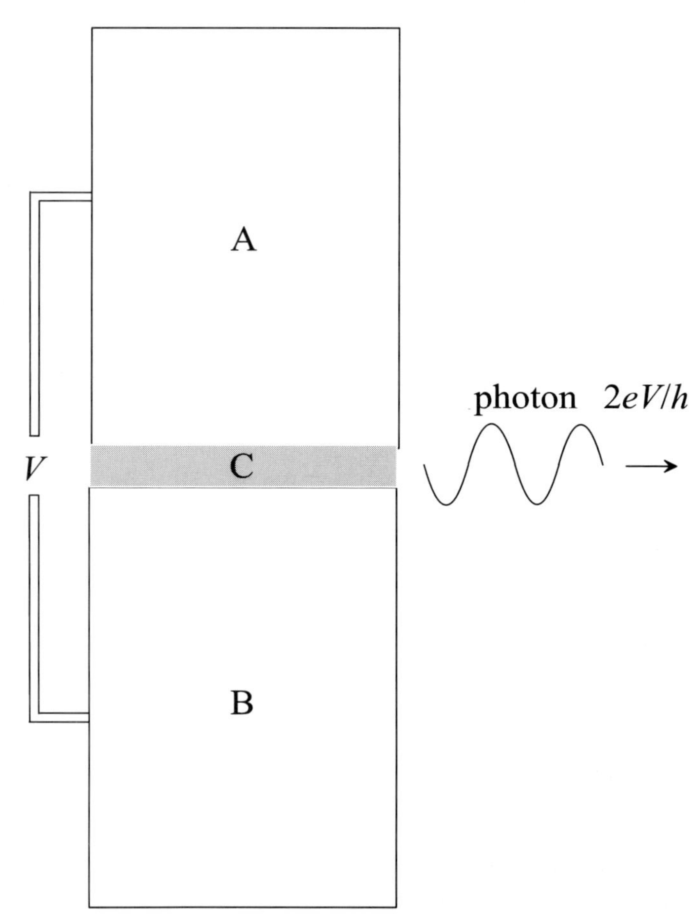

FIGURE XII-10. The Josephson effect. Two superconductors A and B are separated by a thin insulator C. A voltage of V volts is applied between A and B. An oscillating supercurrent begins to flow between A and B through C, and this produces an emission of photons (electromagnetic waves) of frequency equal to $2eV$ divided by the Planck constant h.

them. The layer C is thin enough that some current, even though small, can flow through it. If the rest of the electrical circuitry (not shown in the figure) is suitably set up, then photons (electromagnetic waves) of frequency $f = 2eV/h$, where e is the electronic charge and h is the Planck constant, are emitted by the junction. Given the voltage, the frequency f depends only on two universal constants, and not on which particular superconductor is used: a remarkable result indeed. The effect follows purely from the existence of Cooper pairs (hence the factor $2e$ in the formula for the frequency) and the rules of quantum mechanics (and therefore the factor h).

You can understand how the Josephson effect comes about by recalling some basic results of quantum mechanics. The superconducting state consists of Cooper pairs which all have zero momentum (i.e. no current flowing) or the same non-zero momentum (i.e. some finite supercurrent flowing). This state can be described by a wave function. I do not need to know, for my present purpose, the exact form of this wave function. But I know that there is some part of it which tells me what the energy and velocity (or equivalently the frequency and wavelength) are for each Cooper pair. Both of these latter quantities stand for oscillations of the wave function, the frequency in time and the wavelength in space. I can combine both of these oscillating properties of a wave function in a single quantity called the *phase*, which is simply an angle, as shown in fig. XII-11. The phase has the following properties, which are a consequence of its being a part of the wave function:

1. Its frequency of oscillation f is directly related to the energy E by the quantum mechanical formula

$$E = hf.$$

2. If the phase is different at two points in the material, then there must be a flow of matter between the two points. If this difference is constant with time, then the flow is also constant. But if the phase difference oscillates in time, so does the flow.

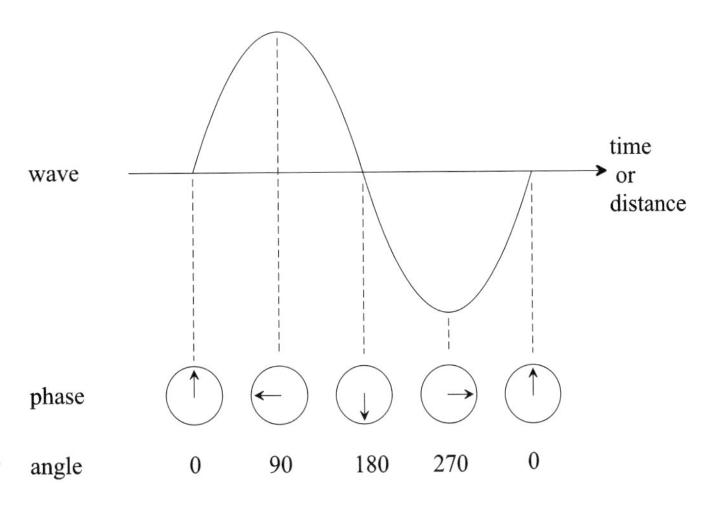

FIGURE XII-11. The phase for a wave function. At the top is one complete cycle of a wave, as it varies with time (relating to its frequency) or at different distances (relating to its wavelength). Now imagine an arrow which is rotating so that it completes one revolution during each cycle of the wave. The direction of the arrow represents the phase of the wave, and five examples are shown for five points on the wave. The magnitude of the phase is given by the angle which the arrow makes with some fixed direction, as shown at the bottom of the figure.

Now we look at what these properties mean for the junction shown in fig. XII-10. There is a voltage difference of V volts between the superconductors A and B on either side of the insulating layer. The potential energy of a charge is equal to the charge multiplied by the voltage it sees. The voltages seen by the Cooper pairs in A and B differ by V volts. Therefore the energy of a Cooper pair, with charge $2e$, differs in the two superconductors by an amount $2eV$. This means that the phase *difference* between the two is oscillating at a frequency f given by $2eV$ divided by h:

$$f = \frac{2eV}{h}.$$

This oscillating phase difference means that there is an oscillating current at the same frequency flowing between the superconductors A and B. Such an oscillating electric current, as we know, is a generator for photons of the same frequency, and this is exactly what is seen to happen. This is the Josephson effect.

One aspect of the Josephson effect is especially striking. A basic feature of quantum mechanics is that the wave function has a frequency which is determined by the energy of the system. This frequency does not, in the many cases I have talked about in this book, make a direct and explicit appearance. But in the Josephson effect, it makes a very palpable appearance, through the frequency of electromagnetic waves (photons) which have been detected and measured in the laboratory. Any lingering doubt one might have about the reality of wave functions should be dispelled by the existence of the Josephson effect.

6 Superconductors at work

The most important practical use of superconductors so far has been to make magnets consisting of spools of superconducting wire carrying an electric current. The wire is made of an alloy of the metals niobium and titanium, and it is superconducting below about 11 K. The magnet itself must be cooled in liquid helium to about 4 K, but the whole equipment is so constructed that the magnetic field can be made available in a volume at room temperature. Such magnets are used for magnetic resonance imaging (MRI) for medical diagnostics, and for experiments in physics and chemistry which need magnetic fields. The MRI technique uses the nuclear magnetic resonance of various nuclei in the human body to give a picture of the body's interior.

Before superconducting magnets became practical and available, magnets consisted of spools of copper wire wound around an iron core. Copper is a good conductor of electricity but still has some electrical resistance. Therefore the passage of a current heated up the spool: electrical energy was being converted into thermal energy. This had two undesirable consequences, namely the need for significant amounts of electrical power (power = rate of energy use), and a limitation on the strength of the fields that could be produced. The need for electrical power to produce a steady magnetic field is especially disappointing, because it should in fact take no energy at all to maintain such a field: a permanent magnet does so with no energy being fed into it. All the energy supplied to a copper-spool magnet gets converted to heat, and the heat has to be somehow carried away in order to prevent a possible meltdown of the magnet.

In a superconducting magnet, the wire has zero resistance and therefore the electrical current produces no thermal energy at all. Because of the very high critical magnetic fields for the alloys used to make this wire, superconducting magnets produce much higher fields than are possible with the more usual copper-spool magnets.

Magnets are by far the most succesful practical application of superconductors to date. There have been proposals, and some experiments, to use such materials in other areas of technology. Examples are superconducting cables to distribute electric power from generating stations, and Josephson junctions used in electronic devices. These efforts continue, but the results till now have not approached the level of superconducting magnets in their availability, reliability and usefulness.

7 Summary

This chapter is concerned with superconductivity, a phase transition which occurs in a large number of metallic elements, alloys and compounds. The basic effect is described in section 1: the total disappearance of electrical resistance below a certain critical temperature.

Section 2 looks at some other properties of superconductors. The superconductivity is destroyed in a magnetic field larger than a certain critical value which is different for different superconductors. They appear to fall into two groups, one with relatively low critical fields and the other with rather high critical fields. The difference lies in the way in which the magnetic field penetrates into the superconductor.

Some clues towards understanding how superconductivity comes about are presented in section 3. The phenomenon is very general, it somehow involves phonons, the electrons appear to act as bound pairs, a finite amount of energy is needed to break up such a pair: these were some of the clues based on experiments done on superconductors.

Section 4 describes how Leon Cooper showed that electrons interacting with one another can form pairs. With this result, John Bardeen, Leon Cooper and Robert Schrieffer used the quantum mechanics of metals with electrons interacting via phonons to explain the phenomenon of superconductivity in all its details.

The Josephson effect is the subject of section 5. Brian Josephson showed that a voltage applied between two superconductors separated by a thin insulating layer produces electromagnetic radiation. One might think that quantum mechanics shows its special features primarily when it is applied to matter on an atomic scale. The Josephson effect is a beautiful example of the universal validity of quantum mechanics: the effect is purely the result of the quantum behaviour of pieces of matter one can see with the naked eye and hold in one's hand.

Section 6 takes up practical applications for superconductors. By far the most successful such application has been in magnets for producing high fields. Some other possible applications, which are not as far along as magnets, are mentioned.

The explanation of superconductivity given by the BCS theory is one of the most impressive accomplishments of the quantum mechanical description of matter as developed in the earlier chapters of this book. The theory in its basic form, as I have outlined it here, involves electrons, phonons, and their mutual interaction. Refinements and extensions of the theory have been made to take account of, among other things, the differences in band structure of different metals, effects due to the presence of magnetic impurities, and effects of electromagnetic radiation. The results of experiments are in every instance well explained by the theory. With the explanation of how superconductivity comes about, we have attained one of the pinnacles of physics.

XIII CONCLUSION

I compared our journey through this book to a mountain ramble. The last chapter, about superconductivity, is like a high peak in the landscape of condensed matter physics. I should like to look back at the terrain we have covered to get there. What I have in mind here is not a summary of the preceding chapters, but rather a review of the key elements necessary for our understanding of the properties of condensed matter. These elements belong to three fundamental parts of physics: quantum mechanics, statistical physics, and electromagnetism. We refer to them as theories, but in a very precise sense. They are supported by vast numbers of experiments, and they predict correctly the results of new experiments. The word *theory* as used here has a significance different from what it has in everyday language as in 'I have a theory about why it always rains on Saturdays.'

Quantum mechanics is the language in physics that we use to describe the world. It incorporates the fact that everything around us is sometimes particle-like, and sometimes wave-like. This basic duality of nature leads to a kind of graininess: just as matter comes in discrete units like electrons and protons and not fractions thereof, quantities like energy and momentum are quantized and cannot take on any arbitrary value we wish. The wave aspect of nature is an expression of a fundamental indeterminism. A simple example of this is an electron contained in a box. If I do a large number of experiments to find out where exactly it is, I find that a pattern emerges. There is a definite probability of finding it in a given region inside the box, and this probability can vary between different regions. What I cannot do is to predict where I shall find the electron the next time I look for it, after I have found out where it is now; herein lies the indeterminism of nature. What is not indeterminate is the *probability* of finding the electron in some specific region, and this probability is predicted correctly by quantum mechanics. Even in considering just one particle, we need to use the idea of probability which is derived from statistics.

We also need statistics in the usual sense of a way of handling large numbers of objects, in order to apply quantum mechanics to matter involving, for example, 10^{21} electrons. We then get a remarkable result, that all particles can be classified into two types, named bosons and fermions, with strikingly different properties. There is another basic result of statistical physics which I merely stated, namely the distribution of particles in different energy ranges at a given temperature. If I fix the total energy and the total number of particles, there are many different ways in which the particles can be distributed in the different energy levels. There is one particular distribution which occurs much more often than any other, and this is the one which actually occurs in nature. The following example illustrates this point. If I toss ten coins repeatedly, I find different numbers of heads and tails each time. There will however be many more times with five heads and five tails than any other possible combination. The larger the number of coins, the more likely will be the result with equal numbers of heads and tails. With the 10^{21} or so particles that we are concerned with in solids, the most probable distribution becomes a practical certainty.

The third and final element in our description of matter around us is the theory of electromagnetism. Nature exhibits four kinds of forces, labelled electromagnetic, nuclear, gravitational, and weak. The electromagnetic force is responsible for the properties that we have considered in this book. The nuclear and weak forces determine the properties of atomic nuclei and their constituents, and the gravitational force becomes dominant at the other end of the scale, with astronomical objects. Quantum mechanics and statistical physics play a fundamental role in all these cases.

Physics as it is today gives a completely satisfactory account of the properties of condensed matter. In the other two areas, namely the nuclear particles and the cosmos, however, there are some observations which cannot be accommodated within the present structure of physics. This is not to say that the physics has failed, but rather that something needs to be added to it so that it can encompass the nuclear and gravitational forces too. A bicycle without a lamp on which I can ride about quite happily during daytime is not a failure because I cannot ride it at night; I just need to mount a lamp on it.

Physics explains why things are the way they are, in the realm of condensed matter. This however does not mean the end of condensed matter physics as an active area of study and research; indeed, far from it. It is like cooking; perhaps not too far-fetched an analogy, for an experimenter's laboratory is somewhat like a kitchen. One often encounters physicists exchanging, or being secretive about, their recipes, and it is not something to eat that they have in mind. The analogy: the basic principles and techniques of cooking are well established, but this by no means spells the end of the art (or science) of cooking. Quite the contrary: there is no end in sight of possible variations on the well-known or the invention of the new, though the basic knowledge and techniques remain the same. So it is in condensed matter physics too. The basic principles, as I have described them in the preceding chapters, are well established, and they give us a good explanation of material properties. But what is not now, and may never be, possible is to predict exactly all the properties of a given material. For example, you might give me a new material, tell me its composition and crystal structure, and ask me if it is going to be superconducting. My answer will be 'Perhaps; but then again, perhaps not.' The reason for such indecision is that one has a very large number of particles interacting with one another, and one can make a theory of its properties only after making approximations. The problem otherwise will be too complicated to solve, even though we know how to formulate it. It could be that a particular approximation causes something to be ignored which was precisely what would have made your new material become a superconductor. One of the skills a physicist needs in this field, as she approximates, is to know what to keep and how much of it, and what to throw away – rather like a cook in his kitchen. A related question is, can we design a material so that it has specified properties, for example a superconducting transition at 300 K? The answer at best can only be 'Not yet.'

The great triumph of condensed matter physics has been its explanation of the properties of things in general: elastic, optical, electrical, magnetic, superconducting and so on. Its task, and its challenge, in the future will be to explain quantitatively and completely the properties of the ever more complicated materials that are being made, and to design new materials which have specified properties. All this still is within the world of inanimate things. Living things are also composed of the same kinds of atoms that make up the inanimate part of the world. Can one then apply the same principles of physics to obtain an understanding of, say, memory, which is comparable to what we already have of superconductivity? It is not clear whether a description that is based purely on physics can be given of all aspects of the living world. There is already progress, however, in analysing some basic biological materials and processes at an atomic level. This field of biophysics, along with condensed matter physics, will continue to be lively areas of study and research. The basic physics is already there. What will be needed as before is artistry in approximating, and the wisdom to know when to stop.

GLOSSARY

absolute zero A temperature of -273.16 degrees on the Celsius scale at which all thermal motion ceases.

alloy A metallic solid containing atoms of more than one element. Common examples are brass which contains copper and zinc, and stainless steel which is composed of iron, nickel and chromium.

ampere The unit for measuring electric current: a flow of 6×10^{18} electrons per second is one ampere.

angstrom A unit of length such that there are a hundred million angstroms in one centimetre; useful in considering lengths on an atomic scale.

angular momentum A quantity which is for a rotating body what linear momentum is for a moving particle. It is equal to the rotational speed multiplied by a quantity which depends on how the mass of the body is distributed around the rotational axis.

annealing Heating a solid, usually an alloy, to an elevated temperature and holding it there for some time. This treatment can change some of the mechanical and other properties of the solid.

antinode The parts of a standing wave where the amplitude of vibration is a maximum.

antisymmetric wave function A wave function which only goes from positive to negative, or vice versa, without changing its magnitude, when two of the particles described by it are interchanged.

atom The smallest entity, a few angstroms in size, which characterises a chemical element. It has a positively charged nucleus and a number of negatively charged electrons, which number determines the element of which it is an atom. The atom itself is electrically neutral, i.e. its total charge is zero.

axis of rotation An imaginary line in a rotating body, around which each part of the body is moving in its own circle. Such an axis for the rotating earth is the line joining the north and south poles.

band A group of very closely spaced quantum energy levels for electrons in a solid which are derived from a single energy level of the isolated atom.

band structure The dispersion curves for the electrons in a solid.

biophysics The study of biological materials and processes using the principles and techniques of physics.

Bloch electron An electron in a crystal whose properties take account of the atomic structure of the crystal.

Bohr magneton A unit, and a fundamental constant, for measuring the strength of the magnetic behaviour of electrons; the magnetic strength of a free electron is very close to one Bohr magneton.

Boltzmann constant One of the fundamental constants of nature, denoted by k_B, particularly important in statistical physics; provides for example the connection between the energy of vibration of the atoms in a solid on the one hand, and the temperature of the solid on the other.

boson The name for particles which obey Bose-Einstein statistics; the wave function for a system of bosons is symmetric, i.e. it is unchanged when two particles are interchanged. Examples are helium atoms, photons and phonons.

Brillouin zone A polyhedron in wave vector space, within which are contained all the possible wave vectors for the electrons in a crystal.

brittleness The tendency of a body to break rather than continue bending when pushed beyond a certain point.

bulk properties The properties of a piece of matter taken as a whole, to be distinguished from the properties of its surface.

c.g.s. units A system of measuring physical quantities in which the units of length, mass and time are centimetre, gram and second respectively.

cc Abbreviation of 'cubic centimetre'; also called a millilitre.

charge A property of matter which leads to the electromagnetic force, just as mass is the property which leads to the gravitational force.

chip A small thin piece of matter, usually made of silicon, on which transistors and other devices are deposited, all on a microscopic scale, to form an electrical circuit.

condensed matter physics The physics of solids and liquids, where the atoms are so close to one another that they cannot be treated as independent particles.

conduction band A band of quantum energy levels in a solid, only partially occupied by electrons. These electrons are capable of carrying an electric current, hence the name.

conduction electrons The electrons in a conduction band.

conservation of energy A fundamental law of nature: the total energy of a system isolated from the outside is always the same, and the only change possible is from one form to another.

conservation of momentum A fundamental law of nature: the total linear momentum and the total angular momentum of a system, isolated from the outside, each remains the same for all time.

continuous spectrum Light has such a spectrum if all frequencies within some range are present in it; sunlight has a continuous spectrum.

Cooper pair The entity found in a superconductor, consisting of two electrons with opposite momenta and spins which can only move as a whole and not independently of each other.

cosmological principle A term borrowed from astronomy; means the arrangement of atoms around an atom in a crystal is exactly the same as around any other atom.

coulomb The unit for measuring electric charge, equal to the total charge of about 6×10^{18} electrons.

critical current The maximum electric current that a superconductor can carry and still remain a superconductor with zero electrical resistance.

crystal A solid whose atoms or molecules are in positions having translational symmetry.

crystal structure The result of placing specific atoms or molecules at the sites of a crystal lattice.

Curie temperature The maximum temperature at which ferromagnetic behaviour is shown; the ferromagnetism disappears above this temperature.

current The term often used when electric current is meant.

current density The current flowing through per square centimetre of the cross-section of a wire.

degeneracy The situation in which more than one wave function corresponds to the same quantum energy level.

density The density of something is the amount of that thing per unit volume, usually taken to be one cc. Thus the density of electrons is the number of electrons per cc. One also says the density of silver to mean the mass of silver per cc of volume. The context makes clear which sense is meant.

diffraction The bending of light or other waves as they pass past an obstacle.

dimensions of space A straight line is one-dimensional: one number specifies every point on it, namely its distance from some fixed point on the line. A flat surface is two-dimensional, because a point on it is fixed by its distances from two perpendicular lines on the surface. Similarly a point in the space around us requires three numbers, its distances from three intersecting planes perpendicular to one another.

diode A device that converts an alternating electric current which changes its direction of flow periodically to direct current, which always flows in the same direction.

dislocation A type of imperfection in a crystal which permits plastic deformation.

dispersion The fact that the speed of light in a substance is different for different frequencies.

dispersion curve A pictorial representation of how the energy (frequency) of a particle (wave) varies with momentum (wave vector). The term is used for particles like electrons and photons as well as for waves like Bloch electrons and light.

doped semiconductor A semiconductor containing a small controlled amount of another element, called the dopant, in order to set its content of electrons or holes.

elastic limit The maximum deformation which an object can tolerate and still show elastic behaviour; if deformed beyond this limit, the object either fractures or shows plastic behaviour.

elasticity The small change in size and shape of an object produced by applying a force to it disappears when the force is removed, leaving the object in exactly the same condition as before.

electric charge Same as charge.

electric field A region of space is said to be permeated by an electric field if an electric charge, an electron for example, feels a force when placed anywhere in it.

electric neutrality The state of an object in which the total positive and negative charges are exactly equal in magnitude.

electromagnet A magnet consisting of a suitably shaped piece of iron or other ferromagnetic material, surrounded by a coil of wire through which an electric current flows. Sometimes, and particularly when the wire is a superconductor, the ferromagnetic material is omitted.

electromagnetic radiation Wave motion of coupled electric and magnetic fields, with a speed in empty space of $c = 3 \times 10^{10}$ cm/sec.

electron One of the fundamental particles in nature, characterized by its mass, negative electric charge, magnetic moment and spin.

electron-volt A unit of energy, equal to the energy gained by an electron in moving across a voltage difference of one volt.

element A substance composed of only one kind of atom, silver for example.

elementary particle physics The physics of the basic interactions among the fundamental particles: electrons, protons, photons, quarks, and so on.

energy An object is said to have energy either because it is moving, or because by some suitable process motion can be produced in it or in another object.

energy level A definite amount of energy that an object can have according to the rules of quantum mechanics.

equation A shorthand way, using symbols, to say that something is equal to something else.

erg The name of the unit in which energy is measured. An object having a mass of two grams and travelling at a speed of one cm per second has a kinetic energy of one erg.

excited state An object, an atom for example, is said to be in an excited state if its energy is greater than the lowest possible value it can have.

exponential notation A compact way, explained on p. 3, of writing very small or very large numbers which are often encountered in physics.

Fermi energy The highest energy which conduction electrons in a metal can have at the absolute zero of temperature.

Fermi momentum The largest momentum which conduction electrons in a metal can have at the absolute zero of temperature.

Fermi surface The surface in wave vector space, formed by connecting the tips of all possible Fermi wave vectors. All quantum states for wave vectors inside this surface are occupied by electrons, and all states outside are unoccupied, at the absolute zero of temperature.

Fermi wave vector A wave vector that corresponds to the quantum state of maximum energy, i.e. the Fermi energy, occupied by an electron in a metal. States of higher energy are unoccupied.

fermion An elementary particle (examples: electron, proton, neutron) to which the Pauli exclusion principle applies.

ferromagnetic material A material in which the magnetic moments of one or more electrons from each atom line up on their own in the same direction even when no magnetic field is present. Examples are iron, nickel and cobalt.

ferromagnetism A phase transformation characterized by the abrupt appearance of permanent magnetism below a certain temperature.

field A scalar or vector quantity whose value depends in general on position and time. Examples are the temperature in the earth's atmosphere (a scalar field), and the electric and magnetic fields in a light wave (vector fields).

force A force applied to an object changes its velocity (in magnitude and/or in direction).

formula A statement that something is equal to something else, usually expressed in terms of algebraic symbols like x, y and so on.

fracture The abrupt breaking into pieces of a material when an attempt is made to deform it.

frequency The number of oscillations per second, often expressed as so many hertz (one hertz = one oscillation per second).

function A quantity f is said to be a function of another quantity x if the value of f depends on the value taken by x. This dependence is expressed by writing the function as $f(x)$. One can also have a function depending on several quantities $x, y, \ldots$ and written $f(x, y, \ldots)$.

fundamental constants Certain unchanging properties that are found in nature:

examples are the mass and electric charge of the electron, the speed of light in free space, the Planck constant, the Boltzmann constant, the Bohr magneton.

gauss The unit in which the strength of a magnetic field is measured. The earth's magnetic field is about half a gauss.

glass fibres Thin strands drawn from glass.

glassy solid A solid in which the atoms or molecules are in disordered positions, without translational symmetry, as also is the case in ordinary glass.

grain Most solids are made up of tiny crystals called grains which are stuck together.

gravitation A basic force between two bodies, derived from their masses.

ground state The quantum state of lowest energy of a system.

hertz One oscillation per second.

high-field (HF) superconductor A superconducting alloy or compound which retains its superconductivity up to magnetic fields which are hundreds or thousands of times larger than those which most superconducting elements like tin or lead can stand before they lose their superconductivity.

hole A fictitious particle, a kind of anti-electron, whose motion replicates the behaviour of a band in which one quantum state is empty and all the others are occupied by electrons.

infinity Suppose x stands for a quantity which continues to grow larger and larger without end. Then one says that x tends to infinity.

insulator A solid which at ordinary temperatures permits no electric current to flow through it when a voltage is applied to it.

interaction The influence which one body exerts on another. The origin of the interaction could be their masses, in which case the interaction is gravitational. If it is their electric charges and magnetic moments, then the interaction is called electromagnetic.

interference The overlapping of two waves to produce a new resulting wave.

Josephson oscillations The oscillating electric current produced between two superconductors which are just barely in contact with each other when a steady voltage is applied between them; also called the Josephson effect.

joule A unit for measuring energy, equal to 10^7 ergs.

kinetic energy The energy possessed by a body because it is moving.

lattice An orderly arrangement of points in space with translational symmetry.

lattice parameters The lengths and angles which are necessary to specify a crystal lattice completely.

lattice site A point in a lattice.

lattice spacing The distance between two adjacent points in a lattice.

line spectrum Light with a line spectrum has only certain frequencies in it, and not others. Light from a neon tube is an example.

linear momentum The mass of a moving object multiplied by its velocity is equal to its linear momentum.

low-field (LF) superconductor A superconductor, usually an element (as distinct from an alloy or compound of more than one element), whose maximum critical magnetic field is less than about a thousand gauss in strength.

macroscopic Term used to denote the scale of everyday things around us.

magnetic domain A portion of a ferromagnetic material within which all the electron magnetic moments are pointing in the same direction.

magnetic field A region of space is said to be permeated by a magnetic field if a magnet, the one associated with an electron for example, feels a torque when placed anywhere in it.

magnetic moment A measure of the strength of a magnet.

magnetic monopole An isolated north (or south) pole of a magnet, existing by itself. It has not so far been found in nature or produced in the laboratory.

mass The property of an object that leads to the gravitational force, and to the fact that a force is necessary in order to change the velocity of the object.

matter Stuff which has mass; the difference between matter and energy has become a bit blurred since Einstein's $E = mc^2$.

mean free path The average distance travelled by an electron or phonon between two successive collisions.

metal A solid in which there are electrons, in number comparable to the number of atoms, free to wander about at all temperatures down to the absolute zero.

microscopic On an atomic scale.

mode of vibration A sound wave of a given frequency and corresponding wave vector that occurs in a crystal; the thermal vibrations of the atoms are equivalent to the sum of many such modes.

molecule Two or more atoms bound to one another and treated as a single entity.

momentum A quantity which combines information about the mass and the motion of a body: linear momentum for translational motion, and angular momentum for rotational motion.

MOSFET Acronym for metal-oxide-semiconductor-field-effect-transistor.

MRI Acronym for magnetic resonance imaging, a technique which uses nuclear magnetic resonance for medical diagnostics.

n-type semiconductor A semiconductor which is so doped that the majority of current carriers are (*n*egative) electrons.

neutrino An elementary particle with no mass or electric charge, but which still has energy and momentum. This state of affairs is allowed by the theory of relativity for particles travelling at the speed of light.

neutron An elementary particle with no charge, and a mass approximately the same as that of the proton.

node The part of a standing wave where the amplitude of vibration is zero.

normal state The state of a superconductor which has lost its superconductivity because it is above the transition temperature, or because it is in a magnetic field larger than the critical field.

nuclear fission The splitting of an atomic nucleus into two or more fragments.

nuclear magnetic resonance The increased absorption of photons, if they have certain frequencies, by the nucleus of an atom placed in a magnetic field.

nuclear magneton A unit, and a fundamental constant, used to express the magnetic moment of neutrons, protons and nuclei; it is about 5.4×10^{-4} times the Bohr magneton.

nucleus What is left of an atom if all its electrons are stripped off. The nucleus is made up of protons and neutrons, contains most of the atom's mass, and is about 10^{-13} cm across.

ohm A unit in which electrical resistance is expressed; one volt applied across a resistance of one ohm causes a current of one ampere to flow through it.

operation The act of doing something to an object, such as rotating it about an axis through an angle.

order A term to denote the symmetries of a crystal.

p-n junction Two different types of semiconductors intimately bonded together across an interface. One is p-type and the other is n-type, meaning that the electric current is predominantly carried by holes and by electrons respectively.

p-type semiconductor A semiconductor which is so doped that the majority of current carriers are (*p*ositive) holes.

particle An idealization of a material object, as something with mass but a vanishingly small size.

Pauli exclusion principle Certain elementary particles (examples: electrons, protons, neutrons) are such that a quantum state cannot be occupied by more than one particle. Such particles are called fermions.

persistent current An electric current that starts going around in a superconducting ring will continue indefinitely without diminution, because it encounters no resistance to its flow.

phase of wave function A wave function oscillates in time as well as in space. The oscillating amplitude at any instant and point can be related to the angle swept by a body rotating at the same rate. This angle is the phase of the wave function.

phase transformation An abrupt change in the structure of a body which occurs at some characteristic temperature (or pressure) and is accompanied by abrupt changes in some of its properties. The melting of ice to form water is an example.

phonon The particle aspect of the quantum description of a solid's thermal energy, whose wave aspect is a sound wave in the solid.

photon The particle aspect of the quantum description of electromagnetic radiation, whose wave aspect is an electromagnetic wave, e.g. visible light.

Planck constant A fundamental constant, denoted by h, which when multiplied by the frequency or the wave vector gives the energy or the momentum respectively.

plastic deformation A change in shape and size of an object produced by an applied force, which persists when the force is removed.

polarization When applied to electromagnetic waves, this means a beam of light, for example, in which the electric field points in the same direction everywhere. When applied to an atom, it means the slight movement of the electrons and the nucleus relative to each other, produced by an electric field.

polarized light A beam of light in which the electric field points in the same direction everywhere.

polarizer Ordinary light passing through a polarizer comes out polarized.

polycrystal This is the usual form of most solids, consisting of tiny crystals called grains which are held together by the forces between atoms from adjacent grains..

position vector Any point in space can be identified by a vector, called its position vector, drawn to it from some fixed point.

positron A particle which is exactly like the electron in all respects except one: its electric charge is positive and equal in magnitude to the charge of the electron.

potential energy The energy that an object has even when it is not moving. Examples are the gravitational energy of a body, and the energy of a charge in an electric field.

precipitate hardening The process of making a metal harder to deform permanently, by annealing so as to produce small clusters of the added atoms distributed through the volume of the metal.

probability The probability of a given result of an experiment, which could give more than one result, is the fraction of a large number of such experiments in which that particular result appears. Thus the probability of heads when I toss a coin is one-half.

proton A particle with a positive charge of the same magnitude as the charge of the electron, and a mass which is about 1840 times the electron's mass.

quantization Quantities like energy and momentum for an object can take only certain discrete values, and nothing in between. This is called the quantization of energy, momentum, etc.

quantum mechanics A description of nature which includes both the particle- and wave-like aspects of matter, and explains its observed properties.

quantum number A number which labels each of the discrete quantum states of energy, momentum, spin, etc.

quantum state Any one of the possible stable states in which an object can find itself, with an energy, a momentum, etc which are specific to that state.

quark The name of the particles which make up protons and neutrons. A proton and a neutron each consists of three quarks.

radiation A term used here interchangeably with electromagnetic waves.

radioactivity A process whereby a nucleus spontaneously ejects one or more particles; an electron, a helium nucleus, a positron, and a photon are among the possibile ejected particles.

reflection symmetry An object has a plane of reflection symmetry if it looks the same when every part of it is imagined to be reflected by the plane acting as a mirror. In two- and one-dimensional objects, it would be a line and a point of reflection symmetry respectively.

refraction The change in direction of an electromagnetic wave, light for example, in going from one medium to another, caused by a difference in speed in the two media.

resistance A term often used when electrical resistance is meant; the property of a conductor whereby passage of an electrical current through it is associated with a voltage across it.

rotation symmetry An object has an axis of rotation symmetry if it looks the same before and after being rotated about this axis through a specified angle.

scalar A quantity which can be specified just by giving its magnitude: the mass of an object, for example.

scattering time The average time between two successive collisions that particles like electrons and phonons in a solid suffer.

Schroedinger equation A fundamental mathematical expression of quantum mechanics. It is used to calculate the wave function, the energy levels and other quantities of interest.

scientific notation The practice of expressing very large and very small numbers as numbers between 1 and 10 multiplied by a power of 10.

second sound The passage of a temperature pulse along a piece of matter with a speed somewhat less than the speed of sound; can occur when the mean free path of the phonons is very long.

semiconductor A solid with a modest energy gap between the highest fully occupied band (the valence band) and the lowest empty band (the conduction band), so that the effective numbers of electrons and holes can be sensitively controlled through temperature and doping.

sintering The process of consolidating fine powder which has been pressed together, by heating it to a high enough temperature without melting it.

solar battery A device made from a semiconductor, which converts sunlight into electricity.

solid state physics The physics of matter which is in the form of solids; also known as condensed matter physics.

solution hardening The process of making a metal harder to deform permanently, by alloying it with other elements.

specific heat The amount of thermal energy needed to raise the temperature of one gram of a substance by one degree Celsius. It is a characteristic property of a given substance.

spectroscope An instrument which enables one to sort out and measure the different frequencies in a beam of light.

standing wave Produced by the combination of two waves of the same wavelength and frequency travelling in opposite directions in the same space. The standing wave has nodes and antinodes which are fixed in space.

state A term which is sometimes used, to mean *quantum state.*

statistical physics The physics of large numbers of particles acting as a whole, the numbers being so large that only the methods of statistics can be meaningfully applied.

statistics A way of handling the features of very large numbers of objects, where one looks not at specific individuals but rather at groups.

superconducting transition temperature The temperature up to which superconductivity exists in a given material. This temperature is a characteristic of the material and is different for different superconductors.

superconductivity A phase transformation characterized by the abrupt and total loss of electrical resistance below a certain temperature.

superconductor An electrically conducting material whose electrical resistivity abruptly vanishes when it is cooled below a certain temperature.

surface properties Those properties of an object which are governed primarily by one layer, or at most a few layers, of atoms at its surface.

symmetric wave function A wave function which is unchanged when two of the particles described by it are interchanged.

symmetry An object is said to show symmetry under a given operation if it appears unchanged after undergoing that operation.

thermal conduction The flow of energy in the form of heat from the hot part to the cold part of a piece of matter.

thermal conductivity The amount of thermal energy flowing per second from one face to the opposite face of a one-cm cube, when the temperature difference between the two is a degree Celsius. It is a characteristic property of a given substance.

thermal energy The energy possessed by an object by virtue of its temperature.

thermal expansion The change in size and shape of an object produced by a change in its temperature.

thermal radiation The radiation given off by a body because of its temperature.

transistor A device made of a semiconductor, often silicon, which is a basic building block in all electronic equipment.

translation The operation of moving something from here to there without rotating it.

translation symmetry This is the symmetry an object has if it looks the same after being moved from here to there through a specified distance without rotation.

travelling wave The usual wave, in which one sees the crests and troughs actually going somewhere.

unit cell A small portion of a crystal lattice, consisting of just enough lattice points to show the symmetry.

unpolarized light A beam of light in which there is no single direction in which the electric field points.

valence band The band of highest energy levels for electrons in a solid in which all the states are occupied by electrons. The next higher band is either empty or only partially occupied by electrons, and is called the conduction band.

variable An algebraic quantity denoted by a symbol like x, to which different values can be assigned.

vector A quantity specified by its magnitude as well as its direction. Velocity and momentum are examples.

velocity The speed of a moving object as well as the direction of its motion are covered by this term. The velocity, having magnitude as well as direction, is a vector.

volt A unit for measuring the potential energy of an electric charge.

watt The unit for measuring power, which is the rate at which energy is used or produced; equal to one joule per second.

wave Something which oscillates, i.e. repeats its shape, at regular intervals in space as well as in time.

wave function A mathematical function that depends on position and time, and embodies the properties of the wave which is associated with a particle or a group of particles.

wave number A quantity obtained by dividing 1 by the wavelength (i.e. taking the reciprocal of the wavelength). It is the same as the number of waves per cm.

wave vector A vector whose magnitude is the wave number and whose direction is the direction in which the wave is travelling.

wave-particle duality The basic aspect of the world around us which leads to quantum mechanics as its proper description: one and the same entity shows wave- or particle-like behaviour depending on the way in which it is observed.

wavelength The distance between two adjacent crests, or two adjacent troughs, in a wave.

Wigner-Seitz cell A polyhedron surrounding an atom in a crystal, of such a shape that every point within it is nearer to this atom than to any neighbouring atom.

work-hardening The increased resistance to plastic deformation which can be produced in a material by a prior process of repeated deformation.

zero-point energy A fundamental consequence of the quantum nature of matter: the lowest possible energy of a system cannot be zero, but must have a non-zero value which is called its zero-point energy.

INDEX